W0091061

SpringerBriefs in Statistics

More information about this series at http://www.springer.com/series/8921

Umberto Cherubini · Fabio Gobbi
Sabrina Mulinacci

Convolution Copula
Econometrics

 Springer

Umberto Cherubini
University of Bologna
Bologna
Italy

Sabrina Mulinacci
University of Bologna
Bologna
Italy

Fabio Gobbi
University of Bologna
Bologna
Italy

ISSN 2191-544X   ISSN 2191-5458   (electronic)
SpringerBriefs in Statistics
ISBN 978-3-319-48014-5   ISBN 978-3-319-48015-2   (eBook)
DOI 10.1007/978-3-319-48015-2

Library of Congress Control Number: 2016955920

© The Author(s) 2016
This work is subject to copyright. All rights are reserved by the Publisher, whether the whole or part of the material is concerned, specifically the rights of translation, reprinting, reuse of illustrations, recitation, broadcasting, reproduction on microfilms or in any other physical way, and transmission or information storage and retrieval, electronic adaptation, computer software, or by similar or dissimilar methodology now known or hereafter developed.
The use of general descriptive names, registered names, trademarks, service marks, etc. in this publication does not imply, even in the absence of a specific statement, that such names are exempt from the relevant protective laws and regulations and therefore free for general use.
The publisher, the authors and the editors are safe to assume that the advice and information in this book are believed to be true and accurate at the date of publication. Neither the publisher nor the authors or the editors give a warranty, express or implied, with respect to the material contained herein or for any errors or omissions that may have been made.

Printed on acid-free paper

This Springer imprint is published by Springer Nature
The registered company is Springer International Publishing AG
The registered company address is: Gewerbestrasse 11, 6330 Cham, Switzerland

Preface

The mainstream of econometric research has been mostly devoted to linear models. Exceptions are mainly due to the need of addressing the issue of non-normality, and it has generally taken the form of assuming random jumps among different regimes describing different linear dynamics and different values of the parameters (typically again from a linear model) in different scenarios, to induce asymmetries in the distribution of variables or to model the tails properly. A notable exception to this general approach is the use of nonparametric tools in the specification of the dynamics of variables, and particularly the use of copula functions. This tool represents a natural way to address the non-normal distribution at the multivariate level, by separating a multivariate distribution in the specification of the marginal distributions and their dependence structure. Actually, copula functions have mostly been used for the study of cross-section dependence and they have become the dominant tool in fields like the analysis of credit risk for portfolios. There has also been a lively stream of applications to the econometrics of time series, even though it has not had the same success as in the risk management field. The aim of this book is to gather the main concepts of copula function theory that can be fruitfully applied to the analysis of time series, and some new ideas, linked to copulas, that represent promising developments. The cornerstone of the applications of copula functions to time series goes back to the beginning of the 1990s and it is the work by Darsow, Nguyen and Olsen, that we call the DNO theorem. The key finding is that there is a one-to-one relationship between copula functions and Markov processes. On the one side, given every Markov process, we can separate the marginal distributions and the temporal dependence structure between the variable at different points in time. On the other side, given a sequence of copula functions and a sequence of marginal distributions we can build a Markov process. Notice that the structure of this finding is parallel to the Sklar theorem, that started the literature on copula functions, and that established a one-to-one relationship between copula functions and multivariate distributions. The Sklar and the DNO theorems also share the same flaw, i.e., some complexity to extend the analysis to arbitrary large dimensions. In both cases, the extension beyond the bivariate application is quite complex, unless for very specific kinds of dependence, mainly elliptical or exchangeable, so imposing relevant limits to high dimension applications. The crucial difference is that the multivariate extension is a mandatory requirement for time series applications, in which clearly it makes no sense at all to restrict the analysis to the bivariate level. This problem gives rise to questions that are very difficult to answer in full generality and that yet are unavoidable for time series applications. For example, if one finds some kind of dependence at the daily level, which dependence should one expect to find at the monthly level? Moreover, if the dependence between two variables of a time series is represented by a bivariate copula, the dependence among all the variables of the time series are represented by the multivariate extension of the same copula? Or, at least, can the multivariate copula be written in analytical form, or can it be only simulated? All these questions have curbed the diffusion of copula applications to time series analysis. These problems have not been solved yet, and the compounded problem of multivariate time series has not been even addressed yet, apart from very isolated contributions. There is another problem that has not been addressed in the literature and is crucial for time series applications: the convolution problem. In the development of a time series, two consecutive variables Y_{t-1} and Y_t are linked together by a specific copula function that emerges from a convolution operator. Namely, the distribution of Y_t is the convolution of the initial level Y_{t-1} and the increment between $t - 1$ and t. In standard econometric applications, the increments are assumed independent, and this makes the distribution of Y_t a convolution in the standard definition of the concept. The definition can be extended to what we have called C-convolution if the initial level and the increment are dependent, where the C term in the definition stands for the copula function representing this dependence. In any case, the dependence structure between Y_{t-1} and Y_t is represented by a copula that is induced by the convolution. These copulas are called "convolution-based" and, differently from standard copula functions, are determined by the distributions of their arguments, i.e., the increments and the initial levels. In this book, we explore the use of convolution-based copulas for the analysis of time series. Because of the complexity of the subject, we limit our analysis to the univariate setting and the standard first-order Markov process, that is the AR(1) model. The plan of our work is the following. In Chap. 1 we show how to use non parametric analysis that are typical of copula function applications to detect nonlinearities in a time series model. In Chap. 2 we go over the main concepts of copula functions, with a focus on the estimation issue. In Chap. 3 we address the DNO approach and the issue of representation and estimation of Markov processes with copula functions. In Chap. 4 we develop our theory of convolution-based copulas in full generality, and we show how to estimate and simulate copulas of this kind. In Chap. 5 we finally present an application to a topic, in which nonlinear and non-Gaussian dynamics has been studied by many authors, i.e., the analysis of the dynamics of the short-term interest rate.

Bologna, Italy   Umberto Cherubini
August 2016   Fabio Gobbi
                                                                                                  Sabrina Mulinacci

Contents

Common Symbols and Notations

$\square$	End of proof
N	The set of natural numbers
$I = [0, 1]$	The unit interval of the real line
$\mathbb{R} = (-\infty, +\infty)$	The real line
$\mathbb{R}^* = [-\infty, +\infty]$	The extended real line
$\mathbb{R}^{*+} = [0, +\infty]$	The non-negative extended real line
$\mathbb{R}^{*+} \backslash \{0\} = (0, +\infty]$	The positive extended real line
$[a, b] \times [c, d]$	Cartesian product of the intervals $[a, b], [c, d]$
$\mathbb{R}^n = \underbrace{(-\infty, +\infty) \times (-\infty, +\infty) \times \cdots \times (-\infty, +\infty)}_{n \text{ times}}$	The $n-$ dimensional Euclidean vector space
$\mathbb{R}^{*n} = \underbrace{[-\infty, +\infty] \times [-\infty, +\infty] \times \cdots \times [-\infty, +\infty]}_{n \text{ times}}$	The $n-$ dimensional extended Euclidean vector space
$I^n = \underbrace{[0, 1] \times [0, 1] \times \cdots \times [0, 1]}_{n \text{ times}}$	Unit cube in $R^n, n \geq 2$
$\mathbf{x} = [x_1 x_2 \ldots x_n]^T$	n-dimensional (column) vector
$\mathbf{x}'$	The transpose of the vector $\mathbf{x}$
C	Copula function
ϕ	Generator of an archimedean copula
$\phi^{[-1]}$	Pseudo-inverse of ϕ
$F(x, y)$	Bivariate distribution function (cumulative probability function) of the random vector $[X, Y]$, computed at (x, y)
F_i	Univariate distribution function (cumulative probability function) of the $i - th$ random variable
$o_p(1)$	Convergence in probability to 0
F_i^{-1}	Generalized inverse of F_i
Ran F	Is its range

C^{∞} — The space of functions $f : R \rightarrow R$ with derivatives of all orders

L^2 — Is the space of random variables with finite first two moments

C^+ — Upper Frechet bound

C^- — Lower Frechet bound (copula for $n = 2$)

$C^{\perp}$ — Product copula

$\bar{C}$ — Survival copula

$H \overset{C}{*} F$ — C-convolution operator

$VaR_{\alpha}(X)$ — Value-at-risk of an exposure X at confidence level α

$1_{\{E\}}$ — Indicator function of the event E

$sgn(x) = \begin{cases} -1 & x < 0 \\ 0 & x = 0 \\ 1 & x > 0 \end{cases}$ — Signum function

wrt — With respect to

rv — Random variable

iff — If and only if

lhs — Left-hand side

rhs — Right-hand side

a.e. — Except than in a set of Lebesgue measure zero

Chapter 1
The Dynamics of Economic Variables

In 1957 Pablo Picasso painted a series of interpretations of an old and famous painting by Velázquez of 1656 called Las Meninas, portraying the court of the Infanta Margarita Teresa. He reinterpreted, partitioned, and distorted the image of the painting in many new images. In this book we are trying the same attempt with a famous model of dynamics used in time series econometrics, and particularly used for the study economic and financial variables. In its simplest form, this is the autoregressive process of order 1, AR(1), denoted as

$$Y_t = \alpha + \beta Y_{t-1} + \epsilon_t, \tag{1.1}$$

where $(Y_t)_t$ is a stochastic process in discrete time, that is a sequence of random variables indexed by the integer-valued time index t, α and β are constant parameters, and $(\epsilon_t)_t$ is a sequence of random variables, providing what is called the set of innovations (or shocks) to the stochastic process.

In standard econometrics textbooks, this basic model is based on the linear representation and assumptions about the distribution of innovations. In this book we attempt an interpretation and extension of time series models like this based on the concepts of copulas and convolution.

Going back to our comparison with Picasso's project, in this chapter we will provide different representations and particulars of the autoregressive model, mainly using graphical representations. All the pictures will be produced with the same set of randomly generated numbers. In other terms, 1000 random numbers were generated only once, turned into standard normal numbers, and used throughout this chapter to provide sequences of innovatios, with different volatility assumptions, and different assumptions concerning the parameters. Then, using this same random material, we will make pictures of our autoregressive models from different perspectives and under different modifications and distortions of the parameters. Our pictures, that will be

© The Author(s) 2016
U. Cherubini et al., *Convolution Copula Econometrics*,
SpringerBriefs in Statistics, DOI 10.1007/978-3-319-48015-2_1

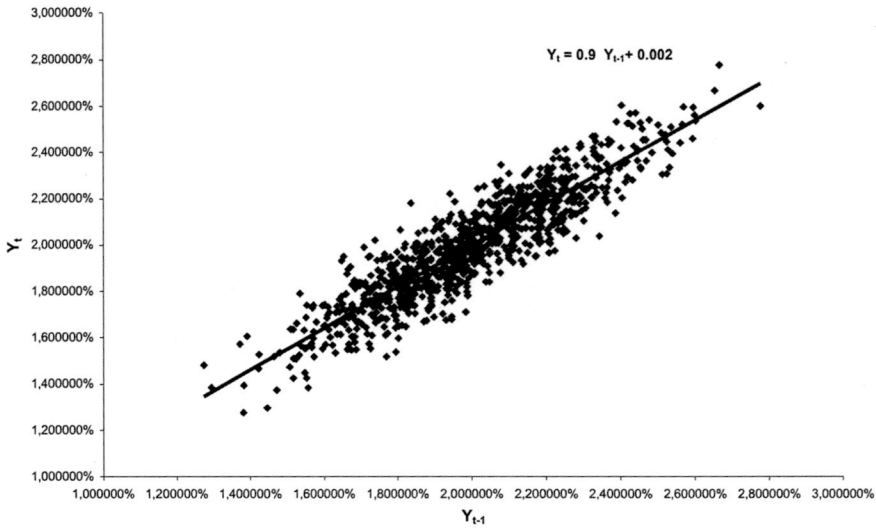

Fig. 1.1 Standard regression scatter plot

far less than the 58 Picasso's remakes of Las Meninas, will have the same target of exploring the inner nature of the classical painting, that in standard econometrics is the regression scatter depicted in Fig. 1.1. We will portrait the model, and its extensions and distortions, beyond the standard representation, using rank scatters, Kendall functions and the like.

While most of the analysis in this chapter will be, just like Picasso and Velazquez, the comparison of two paintings, namely the standard autoregressive models and their distorted, or we could say "cubist," versions, in the end we will introduce real world data. Here we will see that opposite to what happens in paintings, real world data are much closer to the "cubist" vision than to the traditional representation of it offered by the standard linear time series econometrics. This will make clear the main motivation for going on looking through this book, investigating the stochastic processes by non standard tools, namely copula functions and convolution.

1.1   The Standard Linear Autoregressive Model

The philosophy behind the standard autoregressive model in Eq. (1.1), and the vast majority of the models applied in time series econometrics, is to provide a specification of the analytical relationships among the variables constituting the stochastic process. The specification used is the simplest one, that is linear.

As for the main features of this linearity specification, it is well known that a crucial role is played by the autoregressive parameter β. If $\beta < 1$, the system is called "stationary" in the econometric literature (with a somewhat different meaning with respect to the use of the term in the literature of stochastic processes), meaning that the impact of a shock does not remain forever in the history of the process. Put in other terms, the long run behaviour of the process does not depend on initial conditions: one may start the process from any arbitrary value and, if no other shocks occur, the process will eventually converge to the same steady state (or stationary value), that is $\bar{Y} = \alpha/(1 - \beta)$.

The case of β greater than 1 is of no use, since in that case any shock will cause the variable to increase or decrease without bounds, putting the process on a diverging path. Different is the case in which $\beta = 1$. This process is called integrated or nonstationary, or unit root process. In this case every shock remains forever in the history of the process, and its impact never fades away: for this reason the model is also deemed persistent. In this case the parameter α is called "drift," and represents the expected change of the process per unit of time. The case $\alpha = 0$ yields the world-famous random walk model.

Stationary autoregressive processes have been used extensively to model the dynamics of several variables of the financial market. The most famous application is represented by mean-reverting models of the short term interest rates, that corresponds to the Vasicek and Cox–Ingersoll–Ross (CIR) models in continuous time (see Vasicek 1977 and Cox et al. 1985) The mean-reverting assumption was also used as a representation device for the volatility dynamics of stock market returns, that are typically modeled borrowing the same techniques used for interest rates.

On the contrary, mean-reversion evidence has been rarely tested on stock market returns, while the random walk model, first introduced by Bachelier in his thesis at Sorbonne university in the year 1900, was refined in the 1960s to what is called the Efficient Market Hypothesis. Here, the idea of persistence is spelled in terms of unpredictability of future market movements based on current available information.

Finally, the unit root hypothesis reached macroeconomics in the 1980s when Nelson and Plosser in a pioneering paper showed that most macro variables, starting from Gross Domestic Product (GDP), followed the dynamics of a unit root process. They found that a relevant part of the innovation reaching these variables was persistent, and that this part was explaining most of the volatility of the innovations. In the field of macroeconomic variables, this discovery was obviously less welcome than in the financial market, because its meaning was that most of the variation of macroeconomic variables could not be predicted by economic models, with little improvement over the forecast represented by their own current values. A stream of literature then extended these findings to a multivariate setting, investigating whether the same permanent shock was responsible for changes in different processes. In other words, the question was whether a set of different variables were sharing the same integrated processes, so that the system would be called cointegrated. The first technique addressing this problem was due to Granger and Engle.

1.2 Modeling Innovations

If one considers a simple linear model, with constant parameters, as that in Eq. (1.1), the distribution of each variable Y_t in the stochastic process is completely determined by the choice of the distribution of the innovations ϵ_t. A crucial point to remind throughout this book is that this choice is crucial if one wants to endow the process with some useful feature, such as for example requiring that all the variables in the stochastic process Y_t be represented by distributions of the same type. In the first place, in order to enforce this, one would quite naturally require to assume the innovation process to be i.i.d., that is to be composed of independent variables with the same distribution. But this would not be enough: in principle, since the analytical relationship defining the process is linear, by definition only choosing the class of stable distributions for the innovations would maintain the same kind of distribution for the variables in the stochastic process Y_t. In particular, the natural choice would be to assume normal omoschedastic innovations, so that the elements of the stochastic process would be also normal.

It is well known that assuming the normal distribution for the process may be unrealistic for many economic variables, and it is certainly so for financial variables. If this is the case, a linear model with gaussian innovations would not do the job. One natural choice in order to preserve the closure of the model with respect to the distribution of the innovations would be to assume a α-stable distribution with α parameter strictly lower than 2, so that the variance of the process would not be defined. By the definition of stability, this would imply that the stochastic process Y_t would be part of the α-stable family.

An alternative route in order to overcome the gaussian distribution of the variables Y_t is provided by the so-called GARCH processes. This second solution is by far the most common one, particularly for applications to financial markets. The idea is that the variance of the variables is itself a stochastic process, driven by the same innovations. So, in the example of the GARCH(1,1) process, we have

$$h_t = \omega + \gamma h_{t-1} + \delta \epsilon_{t-1}^2, \tag{1.2}$$

where ω, γ and δ are constant parameters, and h_t is the variance of innovation ϵ_t. So, the disturbances in the model are not i.i.d, since the variance of the distribution changes from one period to the other. It may be proved that because of this change in variance, the unconditional distribution of the variable is leptokurtic, that is it exhibits fat tails. As for the dynamics of the variable that represents its second moment, it may also be stationary or integrated, depending on whether the sum $\gamma + \delta$ is strictly lower than or equal to 1. Likewise, also in this case this means that shocks reaching the variance may fade away or remain forever in the history of variance. As of today, the GARCH approach is so common that it is used, as we will see, simply as a preliminary filter to reduce the nonnormality of the process.

For the analysis that follows, we generated two series of innovations, built on the same random draw of standard normal innovations. In one case, we assume i.i.d.

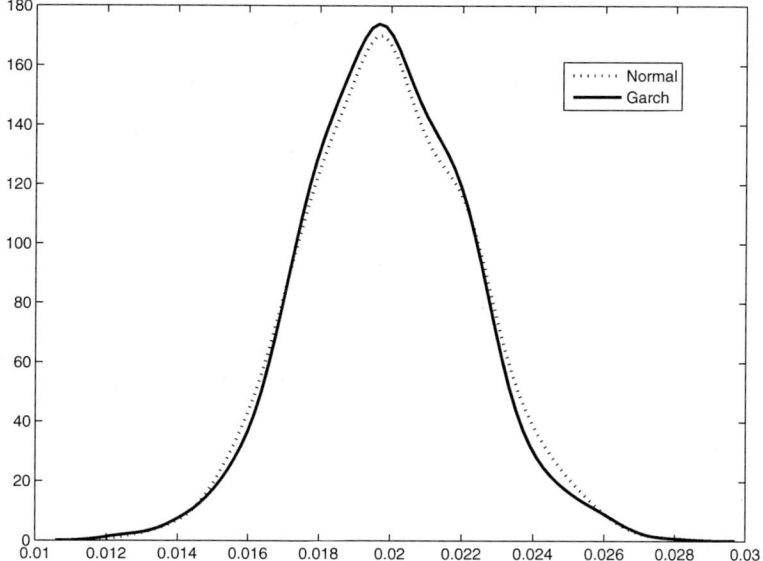

Fig. 1.2 Probability densities of the innovation processes

normal innovations with standard deviation of 0.10% (keeping in mind an application to interest rates). In the other case, we consider a GARCH(1,1) dynamics of volatility as in Eq. (1.2), with $\gamma = 0.7$ and $\delta = 0.1$, which implies a quite high persistence, while the ω parameter was chosen in such a way that the steady state value of the volatility was the same as that chosen for the normal model, that is 0.1%. Figure 1.2 reports the probability density of the two innovation processes.

1.3 A Non-linear Modification of the Model

Here we try a somewhat unusual modification making the autoregressive model non-linear. We may assume that the persistence parameter may change with the process. Generally speaking, one could think of changing Eq. (1.1) as follows

$$Y_t = \alpha + \psi\left(Y_{t-1}\right) Y_{t-1} + \epsilon_t, \tag{1.3}$$

where $\psi(x)$ is a function from the positive line to the unit interval. In this model, the degree of persistence of the shocks changes with different levels of the process. The first example that comes to mind is

$$Y_t = \alpha + \exp\left(-\beta Y_{t-1}\right) Y_{t-1} + \epsilon_t \tag{1.4}$$

in which the degree of persistence decreases exponentially with rising levels of the process. So, if the process were a short term interest rate, this would imply that the mean-reversion force would be higher the higher the level of of the interest rate, while it would become more persistent when the interest rate is near to the zero bound.

A simple extension of this idea would be to define a positive level of the process when the autoregressive coefficient reaches 1, while letting the coefficient to decrease when the process drifts further and further away from this level. An idea to design this dynamics is reported in the equation below

$$Y_t = \alpha + \exp\left(-\beta(Y_{t-1} - \hat{Y})^2\right) Y_{t-1} + \epsilon_t \tag{1.5}$$

Intuitively, the persistence parameter should increase when the level of the process gets closer to a level $\hat{Y}$. From the opposite point of view, the mean-reversion force would get stronger the further the level of the process from level $\hat{Y}$.

In what follows, we will analyze the impact of the two different models above on the distribution of the process, and more to the point of the dependence structure between Y_t and Y_{t-1}. In particular, we will refer to three model specifications. In all cases, we maintain $\alpha = 0, 20\%$ as in the persistent linear model, and in all cases we recover the parameter β in such a way to ensure that 2% be a stationary point. For the rest, the models are defined as follows:

- model 1: Eq. (1.4)
- model 2: Eq. (1.5) with $\hat{Y} = 0$
- model 3: Eq. (1.5) with $\hat{Y} = 1\%$

In practice, model 1 and 2 differ for the linear and quadratic shape of the function in the exponential, while model 2 and 3 differ because in the latter case the unit root is reached in the interior of the domain of the function.

In Fig. 1.3 we report the density function of the process in the three models, compared with the standard linear autoregressive models.

Of course, one can observe that the specifications applied are quite arbitrary and ad hoc. This is exactly the argument that motivates this book. In general nonlinear autoregressive processes of the kind described in Eq. (1.3) are not known in structural form, meaning that we are not given an analytical shape for function $\psi(x)$. In all these cases, when the structural relationship is not known and the linear relationship is not borne out by the data, the only tools available will be those of the dependence analysis described in this book.

1.4 Copula Function Representation

We now provide pictures of the models sketched above using different graphical tools, representing the non parametric techniques available to design the dependence structure. These techniques refer to copula function theory, that enables to study

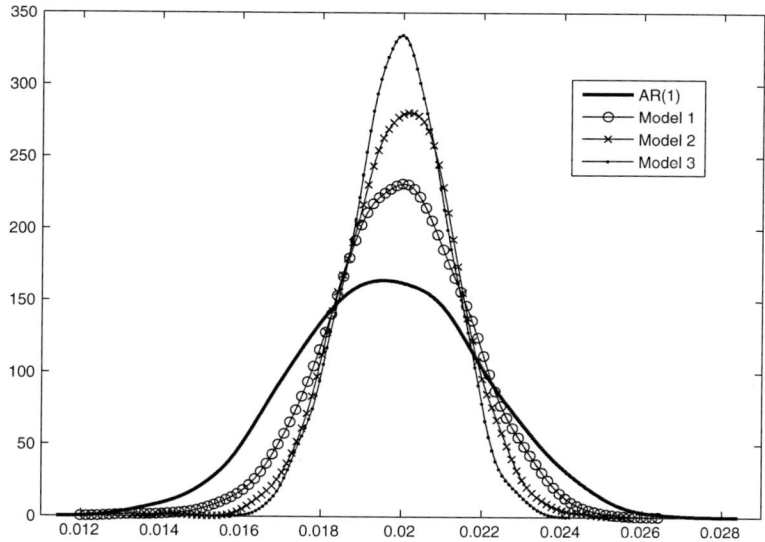

Fig. 1.3 Probability densities of the three models compared with the standard linear autoregressive model

the dependence structure among variables by getting rid of the parametric structure of marginal distributions. Marginals are turned into uniform distributions and the dependence structure is studied among these transformed variables.

Before going through this, we have to discuss another important property of the model specifications used in this chapter, and maintained in most of this book. This is the Markov property. We do not give the formal definition, that will be addressed in future chapters, but we describe the meaning of the concept, and the implication for the copula representation. A stochastic process is endowed with the Markov property if past information of the process does not provide any help over current information in order to forecast its future developments. In other words, the value Y_t is all that is needed to forecast Y_{t+1} and beyond.

Intuitively, the Markov process enables a substantial simplification of the representation of the dependence structure of the process. In fact, in general the copula representation of an arbitrary stochastic process is given by a copula of the same dimension of the process. So, if one wants to represent an n-dimensional stochastic process $\{Y_0, Y_1, \ldots, Y_t, \ldots, Y_n\}$, in general one would need an n-dimensional copula $C(u_0, u_1, \ldots, u_t, \ldots, u_n)$, where u_t denotes the uniform distribution corresponding to the variable Y_t of the process. It is evident that if the process is endowed with the Markov process this representation is redundant and too general. If all one needs to forecast Y_t is information about Y_{t-1} it is clear that the dependence structure of the system can be represented by a sequence of bivariate copula functions $C_t(u_{t-1}, u_t)$. In case one makes the usual assumption that the dependence structure is not time dependent, one could represent all the process by a single copula function C.

The way in which these bivariate copula functions can be combined together to yield the n-dimensional copula representing the system was first analyzed by Darsow et al. (1992), and will be covered in Chap. 3. For what matters here, this provides the justification of why here we describe a whole stochastic process by bivariate representations.

In what follows we portray the bivariate dependence structure of the autoregressive processes described above using two typical techniques of the copula function approach: the rank scatter diagram and the Kendall function representation.

1.4.1 Rank Scatter Diagrams

Since every bivariate relationship can be represented by changing the two marginal distributions, a simple portrait of the dependence function linking the variables can be obtained by transforming the variables into ranks, and reporting the ranks in a scatter diagram.

In Fig. 1.4 we report the scatter diagrams of the linear autoregressive model (1.1). The figure represents rank scatter diagrams of Y_{t-1} and Y_t for three increasing levels of persistence and with normal and GARCH innovations.

Going from top to bottom, the persistence parameter β in Eq. (1.1) increases from 0.5 to 0.9, and finally to 1, that is the unit root model. Notice that the increase of persistence is captured by the increase in dependence, with the scatters that concentrate around the diagonal more and more while the degree of persistence increases. On the left-hand side of the panel we report models with normal innovations, and on the right we have the dependence structure with GARCH innovations. Notice that the diagrams are almost indistinguishable, except maybe for the unit root case. Moreover, it is clear, and can be easily verified, that increasing the size of volatility, which turns into the increase in the shocks, does not change the ranks. So, it seems that in the world of linear autoregressive Markov processes, the degree of persistence could be simply gauged by looking at the rank scatter diagram.

The world of non-linear models is less obvious, and more interesting. If we assume that the data generating process be like the three nonlinear models described above, one would recover the three scatter diagrams reported in Fig. 1.5. Notice that in this case of course the meaning of our scatter is that of an average dependence in the process. In a real-world applications, dependence would change from pair to pair, but we would not be allowed to discover the rule that generates the dependence of each pair, unless we observe the data generating structure. Remember that real world can be even worse: we are lucky here that at least we are allowed to know for sure that the system is Markov. If this were a case of an actual observed process, we could only describe it by this average representation, in the hope that the ergodic structure could ensure that the same average dependence would show up in the future data. In no case would it be possible to trace back the structure to the three nonlinear dynamic models. Figure 1.5 shows that the scatters dispersion increases from model 1 to 2 and 3. This would point out to lower persistence.

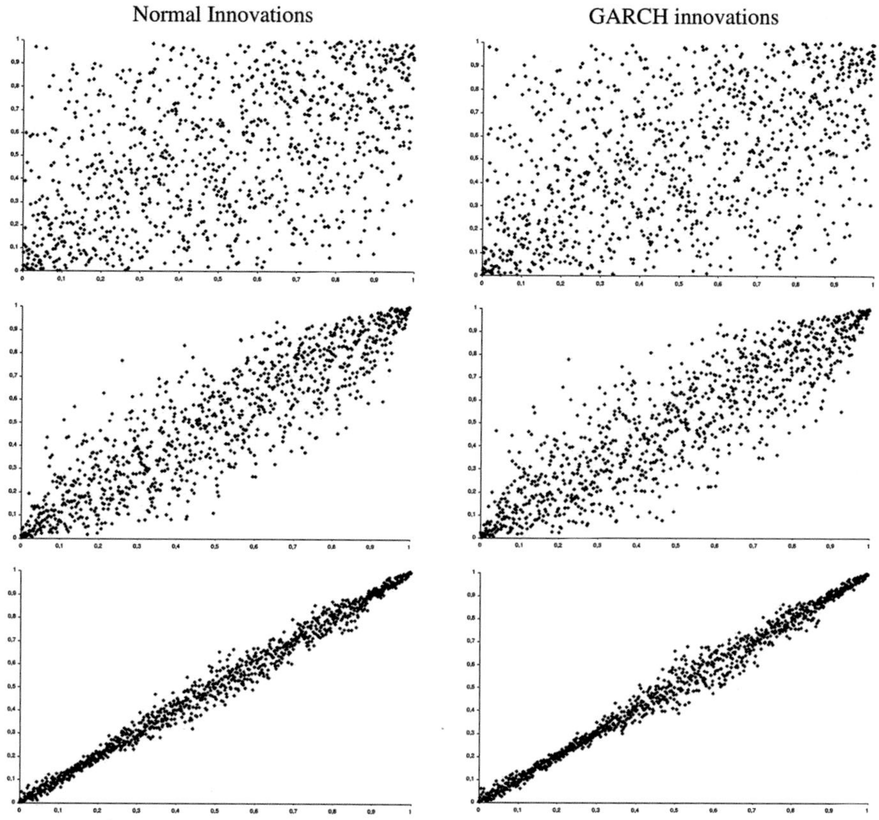

Normal Innovations   GARCH innovations

Fig. 1.4 Scatter diagrams of the linear autoregressive model

Fig. 1.5 Scatter dispersion for the models 1, 2 and 3

1.4.2 Kendall Functions

Another non parametric representation of dependence that is quite common in the copula literature is the so-called Kendall function, which represents the probability associated to the joint distribution, just like the uniform distribution is known to represent the probability of univariate distributions. The empirical Kendall function can be easily retrieved from samples of data, and under some calibration procedures

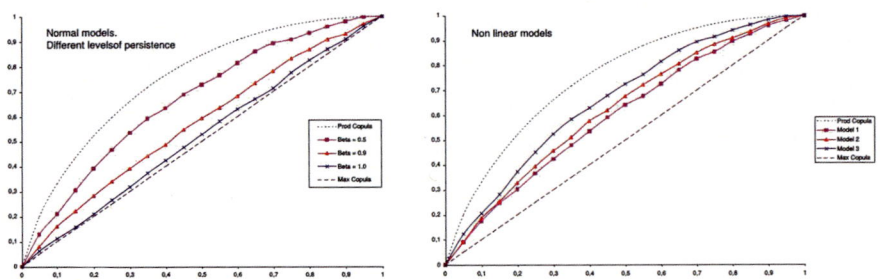

Fig. 1.6 Kendall functions of the dependence relationship between Y_{t-1} and Y_t in the autoregressive model

it is also used to select the copula function that provides the best fit. It also provides a reference graph for the degree of dependence in the data and how it changes across the whole distribution. The extreme cases are perfect dependence, in which the Kendall function coincides with the uniform distribution and represents the diagonal line in the cartesian diagram of the unit square, and the case of independence, that corresponds to a curved line defined by

$$K(x) = x - \ln(x)x \qquad (1.6)$$

In Fig. 1.6 we report the Kendall functions of the dependence relationship between Y_{t-1} and Y_t, in the autoregressive models considered. On the left-hand side, we report the linear models with normal disturbances, and on the right side the three nonlinear models considered. In the linear models picture, the Kendall functions are ordered according to the degree of persistence, with the higher persistence model closer and closer to the diagonal. The Kendall function corresponding to the unit root case is almost indistinguishable from the diagonal. As for the nonlinear models, the Kendall functions are very close together and do not discriminate very much among the three models.

In order to provide further investigation of the models, Kendall functions can be applied to appraise differences in ergodic behavior. In ergodic processes, the dependence between future values of the process and the current level will be weaker for observations further in the future. In other words, there will be a time in the future in which future and current values will be independent. Of course, models can differ for the amount of time needed to reach such independence. Some will need one week, others one month, and some may need even more. Some may need such a very long time that they are hardly distinguishable from integrated processes: they are called *long memory* processes.

In Fig. 1.7 we report the Kendall functions for different time differences of the process. Namely, in each graph we report the Kendall functions of the dependence between the current value of the process and one period, five and twenty periods later. In the idea of application to a daily time series of interest rates or prices of financial markets this can be approximately considered as the dependence on a day, a week

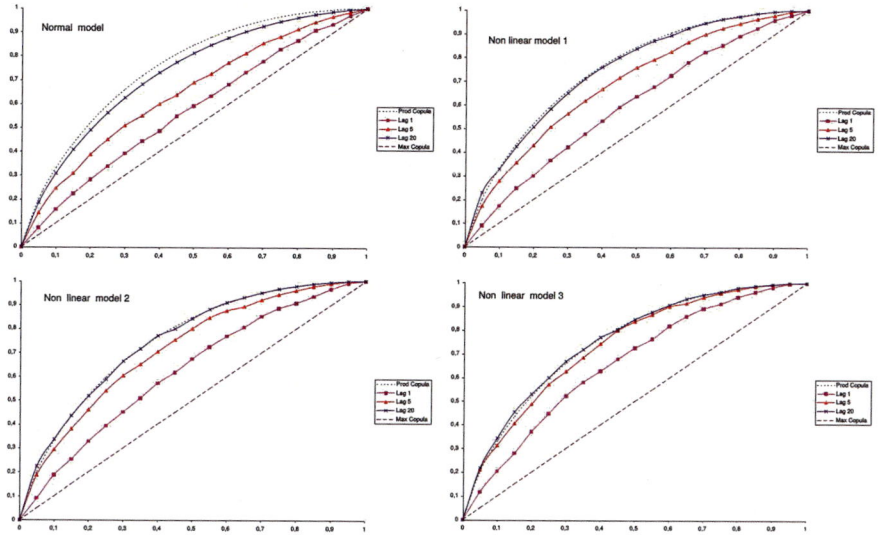

Fig. 1.7 Kendall functions for different time differences of the process

or a month of trading. The four pictures refer to the nonlinear models and the linear model, with high persistence ($\beta = 0.9$), which is reported for comparison. Reading the set of pictures clockwise we may compare the linear model and the nonlinear ones, from model 1 to model 3. Now we see some differences in the nonlinear models. For all of them the Kendall function at one month is undistinguishable from that corresponding to independence. But for all of them the degree of dependence at one week is lower than that one of the linear persistent model. Moreover, the time to reach independence decreases from model 1 to model 3. More precisely, with model 3 dependence evaporates in just one week.

1.5 Convolution Representation

While the copula based representation of Markov processes goes back to the early 1990s, here we propose a new approach, which gives the title to this book, that suggests a strategy to calibrate the copula representation discussed above. The idea is very simple, and is based on a trivial rewriting of the standard autoregressive process in Eq. (1.1):

$$\Delta Y_t = (\beta - 1)Y_{t-1} + \epsilon_t, \tag{1.7}$$

where $\Delta Y_t = Y_t - Y_{t-1}$ is the difference operator. This rewriting in first differences suggests to represent the process in terms of dependence between Y_{t-1} and ΔY_t, that is between levels and increments of the process. Notice that while in the analytical

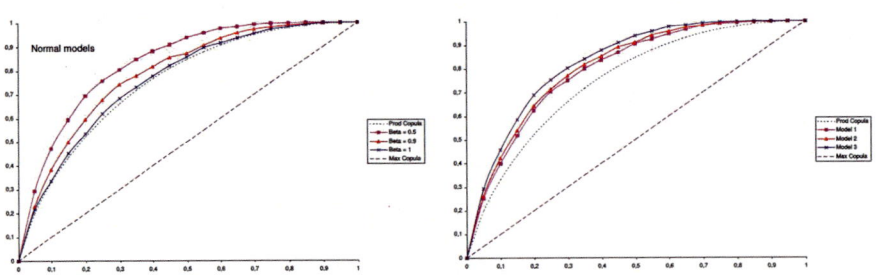

Fig. 1.8 Dependence structure of levels and increments for the linear and non linear models

representation of the model this rewriting makes no difference, in the language of the analysis of dependence it makes a radical difference from the copula representation. While the copula approach builds Markov processes from the dependence structure among levels Y_{t-1} and Y_t, here we start one step behind. We first study the dependence between Y_{t-1} and ΔY_t, and from that (and an assumption about the probability distribution of innovations) we have the problem of recovering the dependence structure between the levels (and the probability distribution of Y_t). So, we somewhat depart from the standard copula approach in which marginal distributions are given and the dependence structure is represented with copulas. Here the distribution of innovations and the copula function linking levels and increments, generate the sequence of dependence structures $C_t(u_{t-1}, u_t)$ and the sequence of marginal distributions of Y_t.

Notice that in the representation of dependence between levels and increments, the implications of persistence is opposite to what happens in the relationship between levels. Namely, now independent increments are evidence of integrated process, while negative dependence structures point out to mean-reverting behavior, with lower persistence the higher the strength of the negative relationship. Figure 1.8 reports the dependence structure for the linear and nonlinear models considered.

1.6 Cointegration

While most of our book will be devoted to univariate time series analysis, it is easy to see how the same techniques based on copulas could be applied to a multivariate setting. In particular, it is easy to see how our Kendall function copula could be applied to verify whether a system of time series is *cointegrated*.

The concept of cointegration was first introduced by Engle and Granger (1987), and it is very easy to define. A set of integrated processes is said to be cointegrated if there exist linear combinations of such variables that are stationary. From another viewpoint, cointegrated means that the same permanent shocks affect the processes in the system, so that one can find opportune combinations of the processes in which the permanent shocks balance each other, and are cleared out.

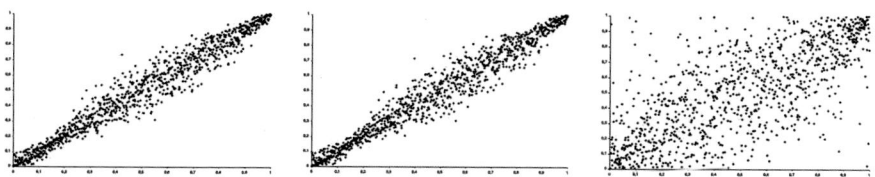

Fig. 1.9 Rank scatter diagram of each variable its lagged variable

In order to make the discussion more precise, without getting into the detailed proof, we may say that a set of n integrated processes may admit r, $r = 1, 2, \ldots$, $n - 1$ cointegrating vectors. If the cointegrated system admits r cointegrating vectors, this means that the system is driven by $n - r$ stochastic trends, that is $n - r$ different common sources of permanent shocks.

While the theory of cointegration is well known, here we are interested in showing how cointegrated systems can be represented and evaluated with our tools of rank scatter plots and Kendall functions. Of course we built our cointegrated system using the same material that we used in the rest of this chapter. Our cointegrating system is composed by 2 integrated time series, with cointegrating system given by $\{1, -1\}$. Again, the inspiration comes from the interest rate dynamics literature, in which some economic theories predict that spreads of interest rates should be stationary.

In Fig. 1.9 we report the rank scatter diagram of each variable and its lagged variable. The left and center panels refer to the two cointegrated series, while the third is related to the so called *cointegration residual*, that is the inner product of the time series and the cointegrating vector. In our case, this is simply the difference of the two variables, or the spread, if the variables were interest rates. Applying the same principles as above, we observe that dependence is very high for the two variables, while is much lower for the residuals.

We know that this means high persistence for the variables and low persistence for the residuals. What we cannot say is whether the variables of the system are integrated and whether the third variable is actually stationary, meaning that it truly is the cointegrating residual of the system. Figure 1.10 provides a check of this using the Kendall function.

As we did before, in Fig. 1.9. we study persistence in the two variables and the cointegration residuals. Namely, we study the dependence of each time series over one, five and twenty day horizons, and we check whether the Kendall functions head toward independence. We see that this is not the case for either of the two variables, for which the degree of dependence decreases by a limited amount. Actually, if we were to measure the dependence over an even longer horizon, the dependence picture would not change any further. Opposite to that, for the residuals dependence decreases rapidly, reaching already almost independence at the five day horizon level.

The behavior of the scatter diagrams and Kendall functions is very easy to understand if we explain how we build our cointegrated system. We simply took the average of the same random walk process and two different low persistence processes. This explains that, as we increase the time horizon, the serial dependence explained by

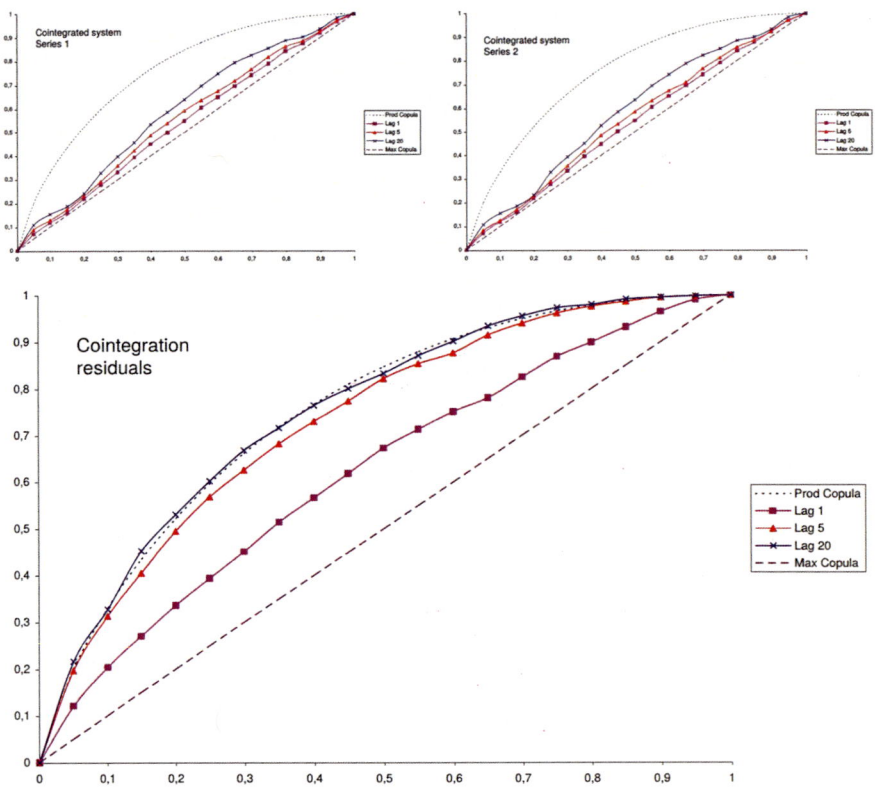

Fig. 1.10 Kendall function and cointegration residuals

the low persistence processes disappears, while the integrated part of the process prevents the serial correlation to decrease beyond that limit. Moreover, it is clear that if the persistent part is erased in the cointegration residuals, the serial correlation quickly fades away as we increase the horizon.

1.7 Higher Order Markov Processes

Here we briefly touch upon the extension of the model to other lags, behind $t - 1$. Even in this case, the extension will not be much addressed in this book. Nevertheless, it may be useful to remind the basic principles of this extension, and show that the same rank analysis could be performed in higher dimensions, even though it would be less straightforward. It is also important, since here we addressed the difference between integrated and stationary processes, how the theory is extended to higher order Markov processes.

The extension is very easily explained if we use the so-called *lag operator L*, defined by $L^j x_t = x_{t-j}$. We can then build a higher order Markov process defining

$$(1 - \beta_1 L)(1 - \beta_2 L) \ldots (1 - \beta_m L) Y_t = \alpha + \epsilon_t \qquad (1.8)$$

Using this factorized representation it is very easy to gauge the requirements for stationarity. Namely, all the parameters β_i, $i = 1, 2, \ldots, m$ must be strictly smaller that 1. If instead a number k of such parameters is equal to 1, we say that the system has k unit roots, or that it is *integrated of order k*.

Of course, when we estimate the system, we do not estimate the β_i parameters directly, but

$$Y_t - \theta_1 Y_{t-1} - \theta_2 Y_{t-2} - \ldots \theta_m Y_{t-m} = \alpha + \epsilon_t, \qquad (1.9)$$

where the parameters θ_i are obtained by computing the product in the factorized version (1.8). Then, the same argument on stationarity versus unit roots can be restated saying that the number of unit roots in the system depend on how many roots of the characteristic equation

$$1 - \theta_1 z - \theta_2 z^2 - \ldots \theta_m z^m = 0 \qquad (1.10)$$

lye on the unit circle.

The meaning of order of the Markov process and integration order is now quite clear. The order of the Markov process is m, meaning that m past observations of the process are needed to forecast its future values. As for the integration order, if we assume that it is 1, it means that only one parameter is equal to 1. Say it is $\beta_1 = 1$. Then, observing that $(1 - L)Y_t = \Delta Y_t$, that is the first difference of the process, we may rewrite the process as

$$(1 - \beta_2 L) \ldots (1 - \beta_m L) \Delta Y_t = \alpha + \epsilon_t \qquad (1.11)$$

and the first difference of the process is stationary by hypothesis. Extending the argument, the meaning of the integration order is quite clear: it is the number of times that we must differentiate the process to make it stationary (Fig. 1.11).

1.8   An Application to Interest Rates Modeling

In this chapter we have introduced a set of techniques to retrieve nonlinearity and nonstationarity in the analysis of time series. This entire book is devoted to techniques that can be used to investigate and estimate these phenomena.

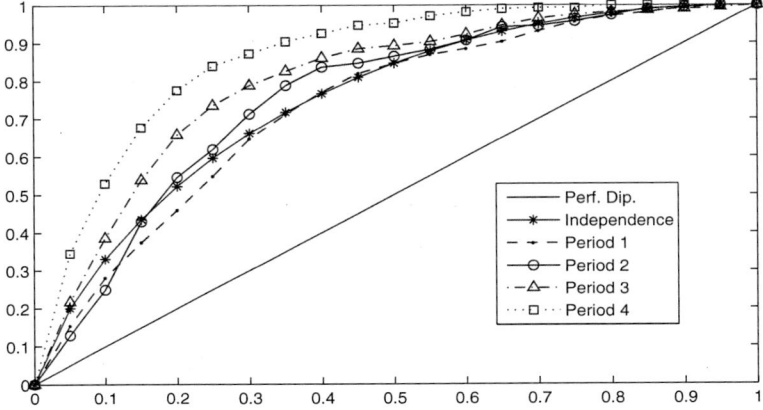

Fig. 1.11 Kendall function analysis of interest rate stationarity

A good training set for our analysis is the dynamics of the interest rates, particularly those for short-term maturities. Short-term interest rates, in fact, are strongly affected by changes in monetary policy, and are subject to structural breaks and changes of regime. Moreover, the problem is ideal for the application of our techniques because the typical dynamics investigated in the literature for this variable is the simple AR(1) model that we are studying in this book.

So, to conclude this chapter, we show a real-world application of one of our tools, namely the Kendall's function between the first difference of the interest rate and its lagged value, that we have called the convolution representation. These are real data, even if we are not disclosing the nature of the data yet. We leave this as a secret until the end of this book, when we will apply our estimation procedures to this data set. For the time being, it is sufficient to disclose that we are dealing with four subperiods of a series of short term rates.

We check that in two cases there is visual evidence of a random walk behaviour, since the increments of the interest rates are independent of its lagged level. In two cases we have instead that the interest rate dynamics is mean reverting, since the empirical Kendall's function lies above the curve denoting independence. Increase of the short term interest rates for this period are negatively associated with the lagged interest rate. This is true across the board, but it looks stronger in the upper tail, that is for higher values of the interest rates. This is actually consistent with evidence that was often reported for variables like this: the mean reversion typically is higher when the interest rates are higher. Our rough first evidence confirms this finding. We wait for the reader at the end of the book to check if there may be more regularities to discover.

References

Cox, J. C., Ingersoll, J., & Ross, S. (1985). A theory of the term structure of interest rates. *Econometrica, 53,* 385–407.

Darsow, W. F., Nguyen, B., & Olsen, E. T. (1992). Copulas and Markov processes. *Illinois Journal of Mathematics, 36,* 600–642.

Engle, R. F., & Granger, C. W. J. (1987). Co-integration and error correction: Representation, estimation, and testing. *Econometrica, 55*(2), 251–276.

Vasicek, O. A. (1977). An equilibrium characterization of the term structure. *Journal of Financial Economics, 5,* 177–188.

Chapter 2
Estimation of Copula Models

2.1 Copula Functions

In this chapter, we introduce copula functions and their main properties. For a more detailed study, we refer the interested reader to Joe (1997), Nelsen (2006) and Durante and Sempi (2015).

Let $(\Omega, \mathcal{F}, \mathbb{P})$ be a probability space and $\mathbf{X} = (X_1, \dots, X_d)$ a random vector there defined with cumulative distribution function F.

It is a known fact that, given a real-valued continuous random variable X, with cumulative distribution function F_X, the random variable $F_X(X)$ is uniformly distributed in $[0, 1]$ ($X \sim U([0, 1])$): $F_X(X)$ is called the *integral transform* of X. If $U \sim U([0, 1])$, then, for every cumulative distribution function F, if

$$F^{\leftarrow}(t) = \inf\{u \in \mathbb{R} : F(u) \geq t\}$$

is the quantile function (or pseudo-inverse) associated with F, then the random variable $F^{\leftarrow}(U)$ is distributed according to F.

The integral transformation of each of the components of the random vector $\mathbf{X}$ brings to the concept of copula functions. In fact, if $\mathbf{X} = (X_1, \dots, X_d)$ so that the random variables X_i have continuous cumulative distribution functions F_i, for $i = 1, \dots, d$, applying the integral transform to each X_i, that is setting $U_i = F_i(X_i)$ for $i = 1, \dots, d$, we get

$$
\begin{aligned}
F(\mathbf{x}) &= \mathbb{P}(X_1 \leq x_1, \dots, X_d \leq x_d) = \\
&= \mathbb{P}(F_1(X_1) \leq F_1(x_1), \dots, F_d(X_d) \leq F_d(x_d)) = \\
&= \mathbb{P}(U_1 \leq F_1(x_1), \dots, U_d \leq F_d(x_d)) = \\
&= C(F_1(x_1), \dots, F_d(x_d))
\end{aligned}
\tag{2.1}
$$

where C is the d-variate cumulative distribution function of $(U_1, \dots, U_d) = (F_1(X_1), \dots, F_d(X_d))$.

© The Author(s) 2016
U. Cherubini et al., *Convolution Copula Econometrics*,
SpringerBriefs in Statistics, DOI 10.1007/978-3-319-48015-2_2

Definition For every $d \geq 2$, a d-dimensional copula is a d-dimensional distribution function whose univariate marginals are uniformly distributed on $[0, 1]$.

Equation (2.1) shows that any multivariate cumulative distribution can be written as a copula having as arguments the marginal cumulative distribution functions. This is formalized by Sklar's theorem:

Theorem 2.1.1 (Sklar 1959)
Let $\mathbf{X} = (X_1, \ldots, X_d)$ *be a random vector with joint cumulative distribution function* F *and univariate marginals* $F_1, F_2, \ldots, F_d$. *Then, there exists a d-dimensional copula function* $C : [0, 1]^d \to [0, 1]$, *such that, for all* $\mathbf{x} = (x_1, \ldots, x_d) \in \mathbb{R}^d$,

$$F(\mathbf{x}) = C(F_1(x_1), F_2(x_2), \ldots, F_d(x_d)).$$

C *is uniquely determined on* $Range(F_1) \times \cdots \times Range(F_d)$ *and, hence, it is unique when* $F_1, \ldots, F_d$ *are continuous.*

From Sklar's theorem it immediately follows that the copula C associated with a joint distribution function F, having continuous marginal distributions, is given by

$$C(\mathbf{u}) = F(F_1^{\leftarrow}(u_1), \ldots, F_d^{\leftarrow}(u_d)).$$

Sklar's theorem allows to decompose a multivariate distribution in its implicit contributors: the marginal distributions and the copula function that links them and that, as a consequence, represents the dependence structure. It gives also a strategy to construct multivariate parametric distributions, starting from the building blocks, by: *(i)* selecting a parametric family for each marginal distribution $F_i(\cdot; \alpha_i)$, for $i = 1, \ldots, d$; *(ii)* selecting a parametric copula family C_θ; *(iii)* combining everything according to Sklar's theorem to obtain the parametric multivariate distribution

$$F(x; \theta, \alpha_1, \ldots, \alpha_d) = C_\theta(F_1(x_1; \alpha_1), \ldots, F_d(x_d; \alpha_d)).$$

The following result is an immediate consequence of Sklar's theorem:

Proposition 2.1.2 (Rank-invariance property) *Let* $\mathbf{X}$ *be a random d-vector with continuous cumulative distribution function* F, *univariate marginals* $F_1, \ldots, F_d$, *and copula* C. *Let, for* $k = 1, \ldots, d$, $g_k : \mathbb{R} \to \mathbb{R}$ *be strictly increasing functions. Then, the random vector,*

$$(g_1(X_1), \ldots, g_d(X_d))$$

admits the same copula C.

Being a copula C a multivariate cumulative distribution function it is said *absolutely continuous* if it admits a density, that is

$$C(\mathbf{u}) = \iint_{[\mathbf{0},\mathbf{u}]} c(\mathbf{t}) d\mathbf{t}$$

for a suitable integrable function $c : [0, 1]^d \to R^+$.

Similarly as for any multivariate cumulative distribution function, in order to guarantee that a generic function $C : [0, 1]^d \to [0, 1]$ is a copula, some suitable condition must be assumed. To analytically characterize copula functions, let us first introduce the concept of C-volume.

For a function $C : [0, 1]^d \to [0, 1]$, the C-volume V_C of the rectangle $[\mathbf{a}, \mathbf{b}] = [a_1, b_1] \times [a_2, b_2] \times \cdots \times [a_d, b_d]$ (with $a_k \leq b_k, k = 1, \ldots, d$) is defined by

$$V_C([\mathbf{a}, \mathbf{b}]) = \sum_{\mathbf{v}} sign(v) C(\mathbf{v}),$$

where the sum is taken over the 2^d vertices $\mathbf{v}$ of the rectangle $[\mathbf{a}, \mathbf{b}]$ and

$$sign(\mathbf{v}) = \begin{cases} 1, & \text{if } v_j = a_j \text{ for an even number of indices,} \\ 1, & \text{if } v_j = a_j \text{ for an odd number of indices.} \end{cases}$$

Theorem 2.1.3 (Characterization of a d-copula)
$C : [0, 1]^d \to [0, 1]$ *is a copula if and only if the following properties hold:*

1. $C(1, \ldots, 1, u_j, 1, \ldots, 1) = u_j$ *for every* $j = 1, \ldots, d$;
2. $C(u_1, \ldots, u_{j-1}, 0, u_{j+1}, \ldots, u_d) = 0$;
3. C *is d-increasing, i.e., the C-volume of every rectangle $[\mathbf{a}, \mathbf{b}] \subset [0, 1]^d$ is non-negative.*

2.1.1 Examples of Copula Functions

The Comonotone Copula
It is the copula
$$M_d(\mathbf{u}) = \min(u_1, u_2, \ldots, u_d), \ \mathbf{u} \in [0, 1]^d$$

and it is the cumulative distribution function of the random vector $\mathbf{U} = (U, \ldots, U)$, with $U \sim U([0, 1])$.

The continuous random variables $X_1, \ldots, X_d$ are called *comonotonic* if they admit as copula M_d: they have the same joint distribution as the random variables $(g_1(W), \ldots, g_d(W))$ for some random variable W and increasing functions $g_k : \mathbb{R} \to \mathbb{R}$, for $k = 1, \ldots, d$.

The Product or Independence Copula
It is the copula
$$\Pi_d(\mathbf{u}) = \prod_{i=1}^{d} u_i, \ \mathbf{u} \in [0, 1]^d$$

and it is the cumulative distribution function of the random vector $\mathbf{U} = (U_1, \ldots, U_d)$, with $U_1, \ldots, U_d$ independent and uniformly distributed on $[0, 1]$.

The 2-Dimensional Countermonotone Copula

It is the copula

$$W_2(u_1, u_2) = \max\{0, u_1 + u_2 - 1\}, \ (u_1, u_2) \in [0, 1]^2$$

and it is the cumulative distribution function of the random vector $\mathbf{U} = (U, 1 - U)$ with $U \sim U([0, 1])$.

Two continuous random variables X_1 and X_2 are called *countermonotonic* if they admit as copula W_2: in this case $X_2 \sim h(X_1)$ for some decreasing function $h : \mathbb{R} \to \mathbb{R}$.

Fréchet Bounds

The following result points out the two bounds between which any copula lies:

Theorem 2.1.4 *For all* $\mathbf{u} \in [0, 1]^d$,

$$W_d(\mathbf{u}) \leq C(\mathbf{u}) \leq M_d(\mathbf{u})$$

where

$$W_d(\mathbf{u}) = \max \left(\sum_{i=1}^{d} u_i - d + 1, 0 \right)$$

The upper bound is the comonotonic copula representing perfect positive dependence. The lower bound W_d is a copula only for $d = 2$. Hence in the case $d = 2$ we have

$$W_2(\mathbf{u}) \leq C(\mathbf{u}) \leq M_2(\mathbf{u})$$

and the bounds represents the perfect negative and positive dependence, respectively.

2.2   Copula Families

2.2.1   The Fréchet Family

This family is obtained through a convex combination (notice that the set of copulas is closed under convex combinations) of the product copula and the comonotonic copula, that is

$$C(\mathbf{u}) = \theta \Pi_d(\mathbf{u}) + (1 - \theta) M_d(\mathbf{u}), \ \mathbf{u} \in [0, 1]^d$$

and $\theta \in [0, 1]$ represents a dependence parameter.

2.2.2 The Farlie–Gumbel–Morgenstern Family

Let $\mathcal{P}^*$ be the class of subsets of $\{1, 2, \ldots, d\}$ with at least two elements. To each $S \in \mathcal{P}^*$ we associate a real number θ_S so that

$$1 + \sum_{S \in \mathcal{P}^*} \theta_S \prod_{j \in S} z_j \geq 0$$

for any $z_j \in \{-1, 1\}$: as a consequence $|\theta_S| \leq 1$, for all $S \in \mathcal{P}^*$.

A Farlie–Gumbel–Morgenstern copula is defined as

$$C(\mathbf{u}) = \prod_{i=1}^{d} u_i \left(1 + \sum_{S \in \mathcal{P}^*} \theta_S \prod_{j \in S} (1 - u_j) \right), \ \mathbf{u} \in [0, 1]^d$$

and the coefficients θ_S represent the dependence parameters.

Any member of this family is absolutely continuous with density

$$c(\mathbf{u}) = 1 + \sum_{S \in \mathcal{P}^*} \theta_S \prod_{j \in S} (1 - 2u_j).$$

In the particular case of $d = 2$

$$C(u, v) = uv + \theta uv(1 - u)(1 - v), \ (u, v) \in [0, 1]^2$$

where $\theta \in [-1, 1]$.

2.2.3 The Archimedean Family

Copulas in this family are represented through a generator which has to satisfy some suitable assumptions (see McNeil (2009)).

We call **Archimedean generator** any decreasing and continuous function $\psi : [0, +\infty) \to [0, 1]$ such that

- $\psi(0) = 1$;
- $\lim_{t \to +\infty} \psi(t) = 0$
- it is strictly decreasing on $[0, \inf\{t : \psi(t) = 0\})$.

ψ is invertible on $[0, \inf\{t : \psi(t) = 0\})$: by convention, we set $\psi^{-1}(0) = \inf\{t \geq 0 : \psi(t) = 0\}$.

An Archimedean generator ψ is said d-monotone if its restriction to $(0, +\infty)$ is d-monotone, that is,

- ψ is differentiable up to the order $d - 2$ in $(0, +\infty)$ and the derivatives satisfy $(-1)^k \psi^{(k)}(t) \geq 0$ for $k \in \{0, 1, \ldots, d - 2\}$ for every $t > 0$,
- $(-1)^{d-2} \psi^{(d-2)}$ is decreasing and convex in $(0, +\infty)$.

A d-dimensional copula C is called Archimedean if it admits the representation

$$C_\psi(\mathbf{u}) = \psi(\psi^{-1}(u_1) + \psi^{-1}(u_2) + \cdots + \psi^{-1}(u_d)), \ \mathbf{u} \in [0,1]^d$$

for some d-monotone Archimedean generator ψ.

Gumbel Copula
It is characterized by the generator $\psi(t) = \exp(-t^{\frac{1}{\theta}})$, for $\theta \geq 1$, from which we get

$$C_\theta(\mathbf{u}) = \exp\left(-\left(\sum_{i=1}^d (-\log u_i)^\theta\right)^{\frac{1}{\theta}}\right), \ \theta \geq 1$$

For $\theta = 1$ we obtain the independence copula as a special case.

Clayton Copula
It is characterized by the generator $\psi_\theta(t) = (\max\{1 + \theta t, 0\})^{-\frac{1}{\theta}}$, for $\theta \in \left[\frac{-1}{d-1}, +\infty\right) \setminus \{0\}$, from which we get

$$C_\theta(\mathbf{u}) = \left(\max\left\{\sum_{i=1}^d u_i^{-\theta} - (d-1), 0\right\}\right)^{-\frac{1}{\theta}}, \ \theta \geq \frac{-1}{d-1}, \theta \neq 0$$

The limiting case $\theta = 0$ corresponds to the independence copula.

Frank Copula
It is characterized by the generator $\psi_\theta(t) = -\frac{1}{\theta}\log(1 - (1 - e^{-\theta})e^{-t})$, $\theta > 0$. from which we get

$$C_\theta(\mathbf{u}) = -\frac{1}{\theta}\log\left(1 + \frac{\prod_{i=1}^d(e^{-\theta u_i} - 1)}{(e^{-\theta} - 1)^{d-1}}\right), \ \theta > 0$$

The limiting case $\theta = 0$ corresponds to the independence copula.
    For $d = 2$, $\psi_\theta(t) = -\frac{1}{\theta}\log(1 - (1 - e^{-\theta})e^{-t})$ is 2-monotone also for $\theta < 0$ and the 2-dimensional Frank copula si defined for all $\theta \in \mathbb{R} \setminus \{0\}$.

2.2.4 The Elliptical Family

A random vector $\mathbf{Y}$ is said to have an elliptical distribution with parameters a d-dimensional vector μ and a positive definite $d \times d$ matrix Σ if

$$\mathbf{Y} \sim \mu + r A \mathbf{U}$$

with $\mathbf{U}$ uniformly distributed on the unit sphere surface of $\mathbb{R}^d$, A given by the Cholesky decomposition of Σ (that is $\Sigma = AA'$) and r a random variable independent of $\mathbf{U}$ with support on $[0, +\infty)$. This distributions admit a density of the form

$$f(\mathbf{x}) = \frac{1}{det(\Sigma)^{-1/2}} g((\mathbf{x} - \boldsymbol{\mu})' \Sigma^{-1} (\mathbf{x} - \boldsymbol{\mu})), \ \mathbf{x} \in \mathbb{R}^{\mathbf{d}}.$$

for some suitable non-negative function g. The normal distribution is easily recovered by taking $g(t) = \frac{1}{(2\pi)^{d/2}} e^{-t/2}$ and the Student's t distribution with m degrees of freedom by taking $g(t) = \frac{\Gamma(\frac{m+t}{2})}{(\pi m)^{d/2} \Gamma(\frac{m}{2})} \left(1 + \frac{t}{m}\right)^{-\frac{m+d}{2}}$. Thanks to Sklar's Theorem, the so called elliptical copula is:

$$C(\mathbf{u}) = \int_{-\infty}^{F_{Y_1}^{-1}(u_1)} \cdots \int_{-\infty}^{F_{Y_d}^{-1}(u_d)} \frac{1}{det(\Sigma)^{-1/2}} g((\mathbf{x} - \boldsymbol{\mu})' \Sigma^{-1}(\mathbf{x} - \boldsymbol{\mu})) \, dx_1 \ldots dx_d.$$

2.3 Dependence Measures

2.3.1 Kendall's Function

We already know that given a continuous random variable X with cumulative distribution function F, its integral transform is the uniformly distributed random variable $F(X)$. The same approach, applied to random vectors of dimensions $d \geq 2$ is less trivial (see for details Nelsen 2006; Genest and Rivest 2001; and Nelsen et al. 2003, among the others).

Let $\mathbf{X}$ be a d-dimensional random vector with cumulative distribution function F and associated copula function C. The *multivariate integral transform* is defined as the random variable

$$W = F(\mathbf{X}) = C(\mathbf{U}) = C(U_1, \ldots, U_d),$$

where U_i is the integral transform of X_i. This random variable is in general no more uniformly distributed and its cumulative distribution function

$$K_C(t) = \mathbb{P}(F(\mathbf{X}) \leq t) = \mathbb{P}(C(\mathbf{U}) \leq t)$$

is called *Kendall's function*.

If $C_1(\mathbf{U}) \geq C_2(\mathbf{u})$ for all $\mathbf{u} \in [0, 1]^d$, then $K_{C_1}(t) \leq K_{C_2}(t)$ for all $t \in [0, 1]$, that is, concordance ordering induces an opposite ordering in the Kendall's function.

Example 2.3.1 (*Fréchet copulas*) Let C_θ be the Fréchet copula

$$C(u, v) = \theta uv + (1 - \theta) \min(u, v), \ (u, v) \in [0, 1]^2, \ \theta \in [0, 1].$$

Then the Kendall's function is

$$K_\theta(t) = t - t \log\{t\theta h_\theta(t) + (1 - \theta)t\} - t \log\{h_\theta(t)\}$$

where $h_\theta(t) = \frac{\theta - 1 + \sqrt{(1-\theta)^2 + 4t\theta}}{2\theta}$.

Example 2.3.2 (Archimedean copulas) Let C be the Archimedean copula given by

$$C_\psi(u, v) = \psi(\psi^{-1}(u) + \psi^{-1}(v)).$$

The Kendall's function is

$$K_\psi(t) = t - \psi^{-1}(t)\psi'(\psi^{-1}(t)).$$

More specifically, we have

- $K_\theta(t) = t - \frac{t}{\theta} \log t$ in the Gumbel case;
- $K_\theta(t) = t + \frac{t^{\theta+1}}{\theta} \left(t^{-\theta} - 1\right)$ in the Clayton case;
- $K_\theta(t) = t - \frac{1 - e^{\theta t}}{\theta e^{-\theta t}} \log \frac{1 - e^{-\theta t}}{1 - e^{-\theta}}$ in the Frank case.

In the $d > 2$-dimensional case (see, for example, Genest and Rivest 2001),

$$K_\psi(t) = t + \sum_{i=1}^{d-1} (-1)^i \frac{(\psi^{-1}(t))^i}{i!} f_{i-1}(t)$$

where $f_i(t) = \frac{d^{i+1}}{dv^{i+1}} \psi(v)$ evaluated at $v = \psi^{-1}(t)$ under the assumption that $\lim_{t \to 0^+} (\psi^{-1}(t))^i f_{i-1}(t) = 0$ for all $i \leq d - 1$.

2.3.2 Kendall's Tau

Roughly speaking, two random variables are concordant if small (large) values of one of the two are likely to be associated with small (large) values of the other.

More precisely, let (x_i, y_i) and (x_j, y_j) be two observations from a vector (X, Y) of continuous random variables. Then, (x_i, y_i) and (x_j, y_j) are *concordant* if $(x_i - x_j)(y_i - y_j) > 0$ and *discordant* if $(x_i - x_j)(y_i - y_j) < 0$.

Given a random sample of observations from a random vector (X, Y), the sample version of the Kendall's tau measure of concordance is given by the difference between the ratio of the concordant pairs and the ratio of discordant pairs. Driven by the same reasoning, the population version of the Kendall's tau is defined as the probability of concordance minus the probability of discordance. More precisely, let (X_1, Y_1) and (X_2, Y_2) be independent and identically distributed random vectors. The population version of Kendall's tau is defined by

$$\tau = P((X_1 - X_2)(Y_1 - Y_2) > 0) - P((X_1 - X_2)(Y_1 - Y_2) < 0).$$

It can be proved that (see Nelsen 2006), if C is the copula of (X_i, Y_i), then

$$\tau = 4 \iint_{[0,1]^2} C(u, v)dC(u, v) - 1 = 4E[C(U, V)] - 1 \tag{2.2}$$

and $\tau = \tau_C$ only depends on the copula C.

Another equivalent expression of the Kendall's tau is given by:

$$\tau_C = 1 - 4 \iint_{[0,1]^2} \frac{\partial C(u, v)}{\partial u} \frac{\partial C(u, v)}{\partial v} dudv.$$

If C is absolutely continuous, then

$$\tau_C = 4 \iint_{[0,1]^2} C(u, v)c(u, v)dudv - 1.$$

It can be trivially checked that $\tau_{M_2} = 1$, $\tau_{\Pi_2} = 0$ and $\tau_{W_2} = -1$ and thanks to the Fréchet bounds $-1 \leq \tau_C \leq 1$ for any bivariate copula C.

It can be easily checked that the Kendall's tau is related to the Kendall's function via

$$\tau_C = 3 - 4 \int_0^1 K_C(t)dt. \tag{2.3}$$

This allows to easily compute the Kendall's tau, when the Kendall's function is know.

Example 2.3.3 Let C_θ be the Fréchet copula of Example 2.3.1. The Kendall's tau for C_θ is given by

$$\tau_\theta = \frac{(1 - \theta)(3 - \theta)}{3}.$$

Example 2.3.4 (*Farlie–Gumbel–Morgestern copulas*) Let C_θ be the Farlie–Gumbel–Morgestern copula

$$C(u, v) = uv + \theta uv(1 - u)(1 - v), \ (u, v) \in [0, 1]^2, \ \theta \in [-1, 1].$$

The value of Kendall's tau for C_θ is given by

$$\tau_\theta = \frac{2}{9}\theta.$$

Example 2.3.5 Let C be the Archimedean copula given in Example 2.3.2. The value of Kendall's tau for C_ψ is given by

$$\tau(\psi) = 1 + 4 \int_0^1 \psi^{-1}(t)\psi'(\psi^{-1}(t))dt.$$

In particular,

- $\tau(\theta) = \frac{\theta-1}{\theta}$ for Gumbel copula and
- $\tau(\theta) = \frac{\theta}{\theta+2}$ for Clayton copula.

Through (2.2) the definition of Kendall's tau can be generalized to higher dimensions, that is (see Joe 1990)

$$\tau_C = \frac{2^d E[C(\mathbf{U})] - 1}{2^{d-1} - 1}.$$

2.3.3 Spearman's Rho

Let $(X_1, Y_1), (X_2, Y_2)$ and (X_3, Y_3) be independent and identically distributed pairs of continuous random variables. The population version of Spearman's rho ρ is defined to be proportional to the probability of the concordance minus the probability of discordance of the two vectors (X_1, Y_1) and (X_2, Y_3). Explicitly

$$\rho = 3\{P((X_1 - X_2)(Y_1 - Y_3) > 0) - P((X_1 - X_2)(Y_1 - Y_3) > 0)\}.$$

It can be easily shown that the Spearman's rho only depends on the copula function C that represents the dependence structure of each vector (X_i, Y_i). In fact, the following equivalent representations of $\rho = \rho_C$ can be easily obtained

$$\rho_C = 12 \iint_{[0,1]^2} C(u, v)dudv - 3 =$$

$$= 12 \iint_{[0,1]^2} uvdC(u, v) - 3 =$$

$$= 12E[UV] - 3 =$$

$$= 12 \iint_{[0,1]^2} (C(u, v) - uv)dudv$$

In particular, considering the representation

$$\rho_C = 12E[UV] - 3,$$

since $E[U] = E[V] = \frac{1}{2}$ and $Var(U) = Var(V) = \frac{1}{12}$

$$\rho_C = 12E[UV] - 3 = \frac{E[UV] - 1/4}{1/12} = \frac{E[UV] - E[U]E[V]}{\sqrt{Var(U)Var(V)}}$$

and Spearman's rho is the Pearson's correlation coefficient of the integral transforms of the original random variables.

Example 2.3.6 Let C_θ be the Fréchet copula of Example 2.3.1, then the value of the Spearman's rho is

$$\rho_\theta = 1 - \theta.$$

Considering the representation

$$\rho_C = 12 \iint_{[0,1]^2} (C(u, v) - uv) du dv$$

and the fact that the Spearman's rho of the comonotone copula is $\rho_{M_2} = 1$, we have

$$\rho_C = \frac{\iint_{[0,1]^2} (C(u, v) - uv) du dv}{\iint_{[0,1]^2} (M_2(u, v) - uv) du dv}$$

that allows to interpret ρ_C as the normalized average difference between the copula C and the independence one (that is Π_2).

Example 2.3.7 Let C_θ be the Farlie–Gumbel–Morgstern copula of Example 2.3.4, then the value of the Spearman's rho is given by

$$\rho_\theta = \frac{\theta}{3}.$$

As shown by the above examples Kendall's tau and Spearman's rho associate different values to the same copula. However, they cannot be too far each other: in fact, they are linked by the following relation

$$-1 \le 3\tau_C - 2\rho_C \le 1.$$

Other relations between them are

$$\frac{1 + \rho_C}{2} \ge \left(\frac{\tau_C + 1}{2}\right)^2 \text{ and } \frac{1 - \rho_C}{2} \ge \left(\frac{1 - \tau_C}{2}\right)^2.$$

Combining all these relations we get the following inequalities

$$\frac{3\tau_C - 1}{2} \le \rho_C \le \frac{1 + 2\tau_C - \tau_C^2}{2}, \quad \tau_C \ge 0$$

and

$$\frac{\tau_C^2 + 2\tau_C - 1}{2} \leq \rho_C \leq \frac{1 + 3\tau_C}{2}, \ \tau_C \leq 0$$

see Theorems 5.1.10 and 5.1.11 and Corollary 5.1.12 in Nelsen (2006).

Starting from the above different representations of the Spearman's rho coefficient, many multivariate extensions have been proposed in literature:

$$\hat{\rho}_C = h(d) \left(\iint_{[0,1]^d} C(\mathbf{u}) d\mathbf{u} - 1 \right), \ \text{see Ruymgaart and Zuijlen (1978)}$$

$$\rho_C^* = h(d) \left(2^d \iint_{[0,1]^d} \Pi_d(\mathbf{u}) dC(\mathbf{u}) - 1 \right), \ \text{see Joe (1990)}$$

$$\bar{\rho}_C = h(2) \left(4 \sum_{k<l} \binom{d}{2}^{-1} \iint_{[0,1]^2} C_{kl}(u,v) du dv - 1 \right), \ \text{see Kendall (1938)}$$

where $h(x) = \frac{x+1}{2^x - 1 - x}$ and C_{kl} is the bivariate marginal copula of C which corresponds to l-th and k-th marginals.

It can be easily proved that

$$\hat{\rho}_C = h(d) \left(2^d \iint_{[0,1]^d} C(\mathbf{u}) d\mathbf{u} - 1 \right) = \frac{\iint_{[0,1]^d} (C(\mathbf{u}) - \Pi_d(\mathbf{u})) \, d\mathbf{u}}{\iint_{[0,1]^d} (M_d(\mathbf{u}) - \Pi_d(\mathbf{u})) \, d\mathbf{u}}.$$

2.3.4 Tail Dependence Parameters

Tail dependence parameters represent a measure of the dependence among two random variables in the upper-right quadrant and in the lower-left quadrant of $[0, 1] \times [0, 1]$. In particular, the *upper tail dependence parameter* λ_U it is defined as

$$\lambda_U = \lim_{u \to 1^-} P\left(V > u | U > u\right) = 2 - \lim_{u \to 1^-} \frac{1 - C(u, u)}{1 - u}$$

while the *lower tail dependence parameter* λ_L as

$$\lambda_L = \lim_{u \to 0^+} P\left(V \leq u | U \leq u\right) = \lim_{u \to 0^+} \frac{C(u, u)}{u}$$

If $\lambda_U \in (0, 1]$ then C has upper tail dependence while if $\lambda_U = 0$ C has no upper tail dependence (the same for λ_L).

As for Archimedean copulas, Clayton copula with parameter $\theta > 0$ has lower tail dependence, since $\lambda_L = 2^{-1/\theta}$, and no upper tail dependence; Gumbel copula has upper tail dependence, since $\lambda_U = 2 - 2^{1/\theta}$, and no lower tail dependence; Frank copula has neither lower neither upper tail dependence. As for the bivariate Student's t copula of Sect. 2.2.4, $\lambda_U = \lambda_L = 2t_{m+1}\left(-\frac{\sqrt{m+1}\sqrt{1-\rho}}{\sqrt{1+\rho}}\right)$, where t_{m+1} is the univariate t-student distribution with $m + 1$ degrees of freedom and ρ is off-diagonal element of Σ.

2.4 Conditional Sampling

In this section, we describe a useful technique in order to generate random pairs from a copula function. We start with elliptical copulas, the Gaussian and the Student copulas, since the simulation technique is very simple. So, let $C(u, v; R)$ be a centered bivariate Gaussian copula with correlation matrix R. The simulation method is given by the following algorithm.

- Find the Cholesky decomposition D of the correlation matrix R.
- Generate 2 independent random numbers $\mathbf{z} = (z_1, z_2)$ from $N(0, 1)$, $(z_1, z_2) \overset{i.i.d.}{\sim} N(0, 1)$.
- Set $\mathbf{x} = D\mathbf{z}$.
- Set $u = \Phi(x_1)$ and $v = \Phi(x_2)$ where Φ is the standard normal distribution.
- (u, v) is the desired pair.

Figure 2.1 shows scatter plots of simulations from a gaussian copula with four different levels of correlations, $\rho = -0.5, 0.1, 0.5, 0.9$ obtained by using this algorithm.

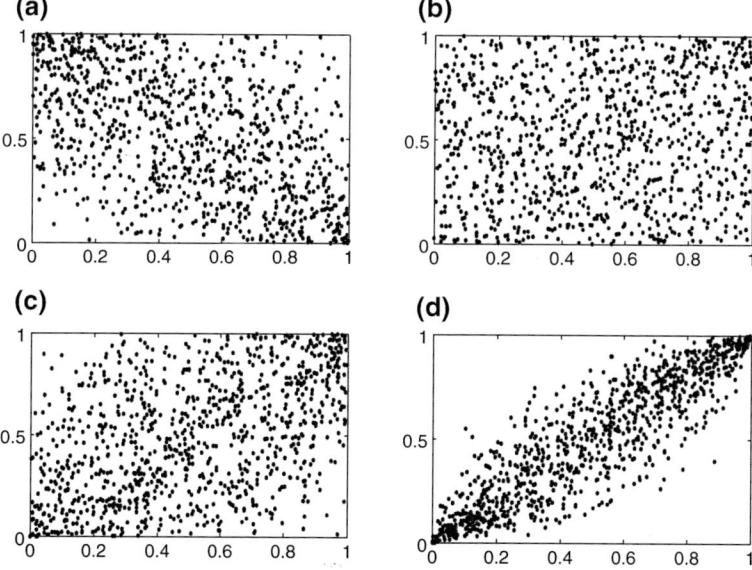

Fig. 2.1 Simulation from a Gaussian copula with **a** $\rho = -0.5$, **b** $\rho = 0.1$, **c** $\rho = 0.5$, **d** $\rho = 0.9$

Analogously, if $C(u, v; R, \nu)$ is a centered bivariate Student copula with correlation matrix R and degrees of freedom ν, the simulation method is very similar to the previous one and algorithm is the following.

• Find the Cholesky decomposition D of the correlation matrix R.
• Generate 2 independent random numbers $\mathbf{z} = (z_1, z_2)$ from $N(0, 1)$, $(z_1, z_2) \overset{i.i.d.}{\sim}$ $N(0, 1)$.
• Generate $s \sim \chi_\nu$ independent of $\mathbf{z}$.
• Set $\mathbf{y} = D\mathbf{z}$.
• Set $\mathbf{x} = D\sqrt{\frac{\nu}{s}}\mathbf{z}$.
• Set $u = T_\nu(x_1)$ and $v = T_\nu(x_2)$ where T_ν denotes the Student's t distribution with d.o.f. ν.
• (u, v) is the desired pair.

Figure 2.2 shows scatter plots of simulations from a Student copula with two different levels of correlations, $\rho = 0.5$ and $\rho = 0.9$ and two different degrees of freedom $\nu = 3$ and $\nu = 30$, obtained by using this algorithm. The parameter ν is a measure of tail dependence. This dependence is higher for small values of ν.

The most useful technique to simulate pairs from an Archimedean copula $C(u, v)$ is the *conditional sampling* method. The method is based on the property that if (U, V) are $U(0, 1)$ distributed r.vs. whose joint distribution is given by C, the conditional distribution of V given $U = u$ is the first partial derivative of C

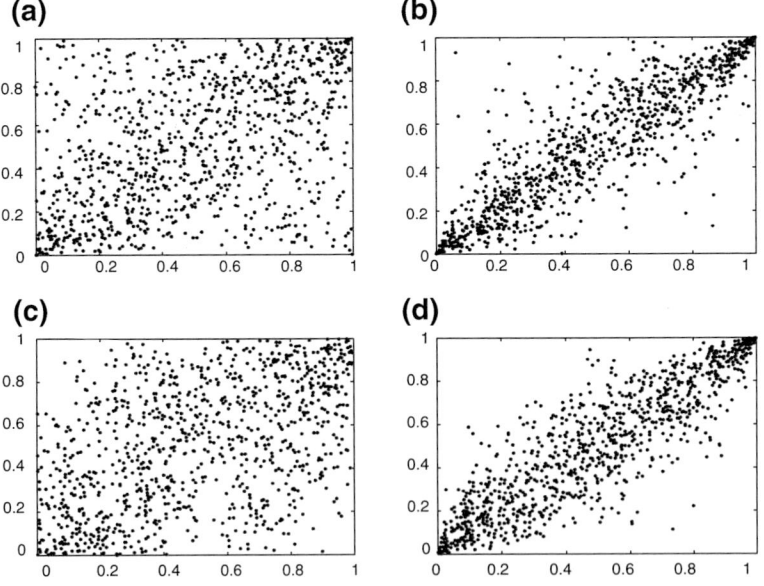

Fig. 2.2 Simulation from a Student copula with **a** $\rho = 0.5$, $\nu = 3$, **b** $\rho = 0.9$, $\nu = 3$, **c** $\rho = 0.5$, $\nu = 30$, **d** $\rho = 0.9$, $\nu = 30$

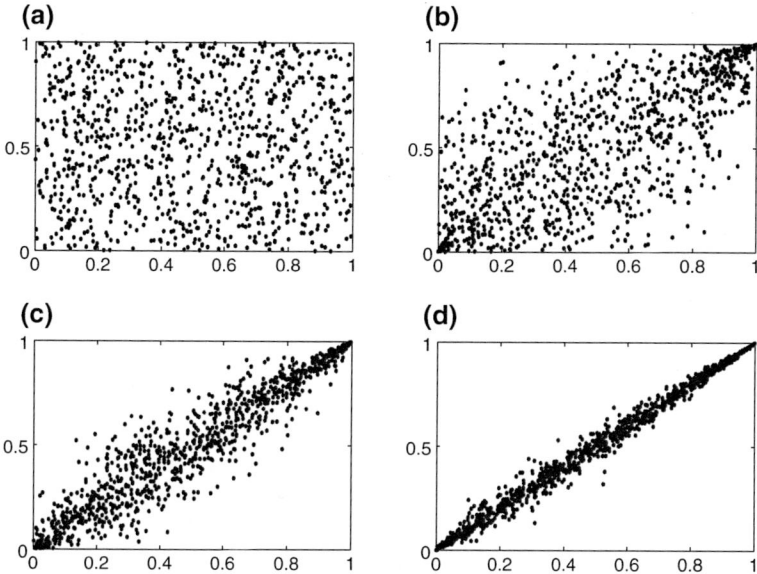

Fig. 2.3 Simulation from a Gumbel copula with **a** $\theta = 1(\tau = 0)$, **b** $\theta = 2(\tau = 50\%)$, **c** $\theta = 5(\tau = 80\%)$, **d** $\theta = 15(\tau = 93\%)$

$$\mathbb{P}(V \leq v|U = u) = D_1 C(u, v) = c_u(v),$$

which is a nondecreasing function of v. With this result in mind the simulation of a pair (u, v) from C is obtained in the following two steps.

1. Generate two independent r.vs. (u, z) from a $U(0, 1)$ distribution: $(u, z) \overset{i.i.d.}{\sim} U(0, 1)$.
2. Compute $v = c_u^{-1}(z)$, where $c_u^{-1}(\cdot)$ is the quasi-inverse function of the first partial derivative of the copula.
3. (u, v) is the desired pair.

Figures 2.3 and 2.4 report scatter plots of simulations from a Gumbel and Clayton copulas obtained by using the conditional sampling algorithm. The Gumbel copula (Fig. 2.3) is a symmetric copula, exhibiting greater dependence both in the upper and lower tails. This effect increases with the parameter value. In particular, the figure shows four different level of dependence expressed in terms of Kendall's τ coefficient (0, 50, 80 and 93%). Figure 2.4 refers to the Clayton copula, which provides an asymmetric dependence, exhibiting greater dependence in the negative tails than in the positive. The figure shows scatter plots of simulations relative to four different levels of dependence expressed in terms of Kendall's τ coefficient (4.75, 33, 72 and 88%).

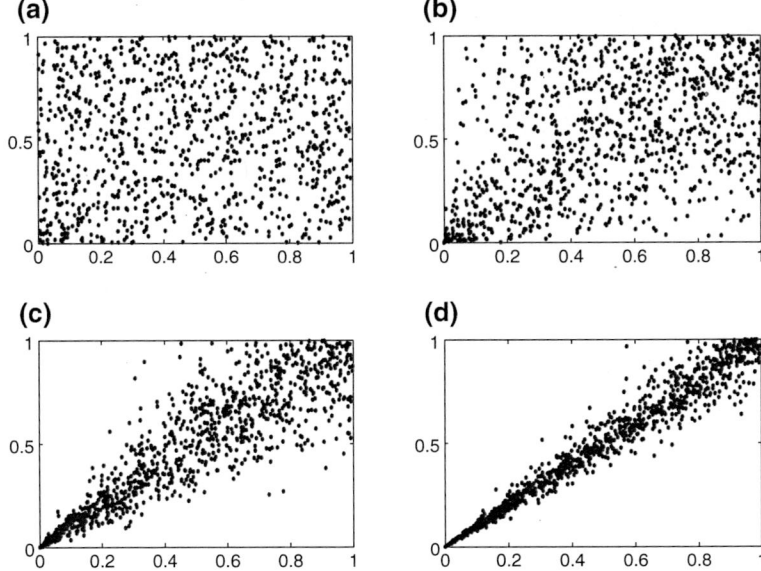

Fig. 2.4 Simulation from a Clayton copula with **a** $\theta = 0.1(\tau = 4.75\%)$, **b** $\theta = 1(\tau = 33\%)$, **c** $\theta = 5(\tau = 72\%)$, **d** $\theta = 15(\tau = 88\%)$

2.5 Inference for Margins Estimation

We briefly present the statistical inference theory applied to copula models. Copulas provide a useful tool to tackle the problem of how to describe a joint distribution because the researcher may deal separately with the needs of marginal and joint distribution modeling. At the same way, from an econometric point of view, a copula model can be estimated in two steps: in the first, one can choose for each data series the marginal distribution that best fits the sample, in the second, one estimates the copula function with desirable properties.

The estimation technique is the maximum likelihood. It is important to remark that in the most cases the maximization procedure requires a numerical optimization of the objective function because a copula is intrinsically a multivariate model and its likelihood involves mixed derivatives.

Let $\mathbf{X} = (X_1, \dots, X_d)$ be a d-dimensional continuous random vector whose joint distribution function is described by the a copula function C and denote by F_k, $k = 1, \dots, d$ the marginal distributions. Therefore, thanks to Sklar's theorem, the joint distribution G of $(X_1, \dots, X_d)$ is

$$G(x_1, \dots, x_d) = C(F_1(x_1), \dots, F_d(x_d)).$$

Now, the likelihood function of the model is obtained from the joint density function of the sample $(x_{1,i}, \ldots, x_{d,i})_{i=1,\ldots,n}$ generated by $(X_1, \ldots, X_d)$. Let g be such a joint density. Since g is the d-th partial derivative of G we get

$$g(x_1, \ldots, x_d) = c(F_1(x_1), \ldots, F_d(x_d)) \prod_{k=1}^{d} f_k(x_k),$$

where c is the copula density and $f_k, k = 1, \ldots, d$, denotes the marginal density of F_k. Suppose that $\phi_k, k = 1, \ldots, d$, represents the vector of parameters of the marginal distribution F_k and θ represents the vector of copula parameters. Therefore, the log-likelihood function of the copula model is

$$\ell((x_{1,i}, \ldots, x_{d,i})_{i=1,\ldots,n}; \phi_1, \ldots, \phi_d, \theta) =$$

$$= \sum_{i=1}^{n} \ln c(F_1(x_{1,i}; \phi_1), \ldots, F_d(x_{d,i}; \phi_d); \theta) + \sum_{i=1}^{n} \ln f_1(x_{1,i}; \phi_1) + \cdots + \sum_{i=1}^{n} \ln f_d(x_{d,i}; \phi_d).$$

This log-likelihood has to be maximized with respect to all parameters $(\phi_1, \ldots, \phi_d, \theta)$ and that could be very computationally intensive especially in the case of a high dimension. To overcome this problem, Joe and Xu (1996) observed that the log-likelihood function is composed by two positive terms: the first involving the copula density and its parameter

$$\ell_c((x_{1,i}, \ldots, x_{d,i})_{i=1,\ldots,n}; \phi_1, \ldots, \phi_d, \theta) = \sum_{i=1}^{n} \ln c(F_1(x_{1,i}; \phi_1), \ldots, F_d(x_{d,i}; \phi_d); \theta)$$

and the second involving the margins and their parameters

$$\ell_k((x_{k,i})_{i=1,\ldots,n}; \phi_k) = \sum_{i=1}^{n} \ln f_k(x_{k,i}; \phi_k), \quad k = 1, \ldots, d.$$

So, they proposed a new estimation technique consisting in two steps:

1. firstly, we estimate the parameters of the margins $\phi_k, k = 1, \ldots, d$ by maximum likelihood

$$\hat{\phi}_{k,n} = \arg \max_{\Phi_k} \ell_k((x_{k,i})_{i=1,\ldots,n}; \phi_k),$$

   where Φ_k is the marginal parameter space;

2. second, given $\hat{\phi}_{1,n}, \ldots, \hat{\phi}_{d,n}$, we estimate the copula parameter θ by maximizing the copula log-likelihood

$$\hat{\theta}_n = \arg \max_{\Theta} \ell_c((x_{1,i}, \ldots, x_{d,i})_{i=1,\ldots,n}; \hat{\phi}_{1,n}, \ldots, \hat{\phi}_{d,n}, \theta),$$

   where Θ is the copula parameter space.

This estimation technique is called inference for the margins (IFM) and the vector $\hat{\eta}_n = (\hat{\phi}_{1,n}, \ldots, \hat{\phi}_{d,n}, \hat{\theta}_n)$ is the IFM estimator. Joe (1997) proves that under some regularity conditions the IFM estimator is asymptotically gaussian and more specifically

$$\sqrt{n}(\hat{\eta}_n - \eta_0) \xrightarrow{d} N(0, \mathcal{G}^{-1}(\eta_0)),$$

where η_0 is the true parameters vector and $\mathcal{G}(\eta_0)$ is known as Godambe information matrix. $\mathcal{G}(\eta_0)$ is given by

$$\mathcal{G}(\eta_0) = A^{-1}V(A^{-1})^T,$$

where $A = \mathbb{E}\left[\frac{\partial s(\eta)}{\partial \eta}\right]$ and $V = \mathbb{E}[s(\eta)(s(\eta))^T]$ where $s(\eta)$ is the score function $s(\eta) = \left(\frac{\partial \ell_1}{\partial \phi_1}, \ldots, \frac{\partial \ell_d}{\partial \phi_d}, \frac{\partial \ell_c}{\partial \theta}\right)$.

2.6 Copulas and Time Series

The estimation technique introduced in the previous section is devoted to applications where the data could be considered independent and identically distributed. However, this assumption has to be rejected for almost every economic time series where the sample are realization of stochastic processes. In this framework the estimation of parameters of the model is obtained by the maximization of the quasi-likelihood functions. Before introducing this methodology, the concept of conditional copula has to be presented.

2.6.1 Conditional Copula

The conditional copula was introduced by Patton (2001, 2006a, b) in order to handle conditioning variables in the analysis of time-varying conditional dependence and for multivariate density modeling. The theoretical framework is the following: X and Y are the variables of interest and Z is the conditioning variable. Let F_{XYZ} be the distribution of the random vector (X, Y, Z), $F_{XY|Z}$ be the conditional distribution of (X, Y) given Z and $F_{X|Z}$ and $F_{Y|Z}$ be the conditional marginal distributions of X given Z and of Y given Z, respectively. The conditional bivariate distribution of (X, Y) given Z can be derived from the unconditional joint distribution of (X, Y, Z) as follows

$$F_{XY|Z}(x, y|z) = f_Z(z)^{-1}\frac{\partial F_{XYZ}(x, y, z)}{\partial z},$$

where f_Z is the unconditional density of Z. The conditional copula of (X, Y) given Z cannot be derived from the unconditional copula of (X, Y, Z): further information is required. Patton defines the conditional copula as follows:

Definition The conditional copula of (X, Y) given $Z = z$, where X given $Z = z$ has distribution $F_{X|Z}(\cdot|z)$ and Y given $Z = z$ has distribution $F_{Y|Z=z}(\cdot|z)$, is the conditional distribution function of $U \equiv F_{X|Z}(X|z)$ and $V \equiv F_{Y|Z}(Y|z)$ given $Z = z$.

Let $\mathcal{Z}$ be the support of Z. So, an extension of Sklar's theorem for conditional distributions shows that a conditional copula has the properties of an unconditional copula for each $z \in \mathcal{Z}$.

Theorem 2.6.1 *Let $F_{X|Z}(\cdot|z)$ be the conditional distribution of X given $Z = z$, $F_{Y|Z}(\cdot|z)$ be the conditional distribution of $Y|Z = z$, $F_{XY|Z}(\cdot, \cdot|z)$ be the joint conditional distribution of (X, Y) given $Z = z$ and $\mathcal{Z}$ be the support of Z. Assume that $F_{X|Z}(\cdot|z)$ and $F_{Y|Z}(\cdot|z)$ are continuous in x and y for all $z \in \mathcal{Z}$. Then, there exists a unique conditional copula $C(\cdot|z)$ such that*

$$F_{XY|Z}(x, y|z) = C(F_{X|Z}(x|z), F_{Y|Z}(y|z)|z), \quad \forall (x, y) \in \mathbb{R}^* \times \mathbb{R}^*, \, \forall z \in \mathcal{Z}. \quad (2.4)$$

Conversely, if we let $F_{X|Z}(\cdot|z)$ be the conditional distribution of X given $Z = z$, $F_{Y|Z}(\cdot|z)$ be the conditional distribution of Y given $Z = z$ and $\{C(\cdot, \cdot|z)\}$ be a family of conditional copulas that is measurable in z, then the function $F_{XY|Z}(\cdot, \cdot|z)$ defined by Eq. (2.4) is a conditional bivariate distribution function with conditional marginal distributions $F_{X|Z}(\cdot|z)$ and $F_{Y|Z}(\cdot|z)$.

The key point in this 'conditional' version of Sklar's theorem is that the conditioning variable Z must be the same for both marginal distributions and the copula and this is a fundamental condition.

Thanks to the notion of conditional copula we can address the problem of estimation of copula-based multivariate time series. From a general point of view, when considering copula-based models for multivariate time series we assume that the marginal distributions are of the form (Patton 2012)

$$X_{kt} = \mu_k(\mathbf{Z}_{t-1}; \eta) + \sigma_k(\mathbf{Z}_{t-1}; \eta)\varepsilon_{kt}, \quad k = 1, \ldots, d,$$

where

$$\mathbf{Z}_{t-1} \in \mathcal{F}_{t-1}, \quad \varepsilon_{kt}|\mathcal{F}_{t-1} \sim F_{kt}$$
$$(\varepsilon_{1t}, \ldots, \varepsilon_{dt})|\mathcal{F}_{t-1} \sim C_t(F_{1t}, \ldots, F_{dt}),$$

where C_t is a d-dimensional conditional copula. In practice, we will allow each series to have potentially time-varying conditional mean and variance each parametrically modeled and we will assume that the standardized residuals ε_{kt} has a conditional distribution F_{kt} which may be both parametric and nonparametric. In the parametric case, it may be modeled as time-varying and its parameters will become part of the vector η.

The hypothesis of time-varying was confirmed in recent econometric literature which contains a preponderance of evidence that the conditional volatility of economic time series changes through the time, see, among others, Andersen et al.

(2006). This motivate the researchers to consider whether the conditional dependence structure also varies through time (see, for example, Rémillard 2010 which considers a test for a one-time change in the conditional copula).

2.6.2 Multi-stage Quasi-maximum Likelihood Estimation

When all components of the copula-based multivariate model are parametric, the most efficient estimation method is maximum likelihood. Under regularity conditions, see White (1994) among others, standard results for parametric time series models may be used to show that the maximum likelihood estimator $\hat{\eta}^n$ is $\sqrt{n}$-consistent and asymptotically gaussian. The drawback of this standard approach is that the number of parameters to be estimated simultaneously can be large, in particular, when the model is of high dimension. To overcome this computational problem it is available an alternative approach which consists in estimating the model in more stages. This requires that the parameters vector can be separated in parameters for the d margins and the conditional copula. This condition is satisfied for the most models used in practice. The estimation procedure consists in a first stage where one can estimate the parameters of the marginal distributions (more precisely we have d stages for a d dimensional model) and a second stage which consists in estimating the copula parameter conditioning on the estimated marginal distributions parameters. This estimation procedure is known as multistage maximum likelihood estimation. White (1994) proves the asymptotic properties of this estimator. In this paragraph, we present a brief summary of these results. From an econometric point of view time series are realizations of a stochastic process on some suitable probability space. Formally, the observed data is generated by the (multivariate) stochastic process $\mathbf{X} = \{(X_{1t}, \ldots, X_{dt}) : \Omega \to \mathbb{R}^d, t = 1, 2, \ldots\}$ on a complete probability space $(\Omega, \mathcal{F}, \mathcal{F}_t, \mathbb{P})$, where $\mathcal{F}_t$ denote the filtration of the model. In a similar framework, the joint densities of a time series have to be considered conditional to $\mathcal{F}_{t-1}$ and the likelihood functions are "quasi" likelihood. For more details the reader may consult the book of White (1994).

In a copula model for time series the data generating process contains a conditional copula and d conditional marginal distributions. Let C_t be the conditional copula of $(X_{1t}, \ldots, X_{dt})|\mathcal{F}_{t-1}$ and F_{kt} be the conditional marginal distribution of $X_{kt}|\mathcal{F}_{t-1}$. So by the conditional version of Sklar's theorem the joint distribution of $(X_{1t}, \ldots, X_{dt})|\mathcal{F}_{t-1}$ is given by $G_t(x_{1t}, \ldots, x_{dt}) = C_t(F_{1t}(x_1t), \ldots, F_{dt}(x_{dt}))$. Suppose that $\phi_k, k = 1, \ldots, d$, is the vector of parameters of F_{kt} and θ is the vector of parameters of the conditional copula. Therefore, the parameter of the model is the vector $\eta = (\phi_1, \ldots, \phi_d, \theta)$ and the joint density function of $(X_{1t}, \ldots, X_{dt})|\mathcal{F}_{t-1}$ is

$$g_t(x_{1t}, \ldots, x_{dt}; \eta) = c_t(F_{1t}(x_{1t}; \phi_1), \ldots, F_{dt}(x_{dt}; \phi_d); \theta) \prod_{k=1}^{d} f_{kt}(x_{kt}; \phi_k).$$

Now, let $x_k^n = (x_{kt})_{t=1,\ldots,n}$ be the time series generated by the k-th component X_{kt} with $k = 1, \ldots, n$. The quasi log-likelihood function of the model, denoted by ℓ^n, is then

$$\ell^n((x_1^n, \ldots, x_d^n)); \eta) =$$

$$= \sum_{t=1}^n \ln c_t(F_{1t}(x_{1t}; \phi_1), \ldots, F_{dt}(x_{dt}; \phi_d); \theta) + \sum_{t=1}^n \ln f_{1t}(x_{1t}; \phi_1) + \cdots + \sum_{t=1}^n \ln f_{dt}(x_{dt}; \phi_d).$$

In the same spirit of the IFM estimation we proceed to the estimation in $d + 1$ steps because the quasi log-likelihood function is obtained by the sum of d marginal quasi-likelihoods

$$\ell_k^n((x_k^n; \phi_k) = \sum_{t=1}^n \ln f_{kt}(x_{kt}; \phi_k), \quad k = 1, \ldots, d$$

and a copula quasi-likelihood

$$\ell_c^n((x_1^n, \ldots, x_d^n)); \eta) = \sum_{t=1}^n \ln c(F_{1t}(x_{1t}; \phi_1), \ldots, F_{dt}(x_{dt}; \phi_d); \theta).$$

The multistage quasi-maximum likelihood estimator (MSQMLE) is the vector $\hat{\eta}^n = (\hat{\phi}_1^n, \ldots, \hat{\phi}_d^n, \hat{\theta}^n)$ whose components are

$$\hat{\phi}_k^n = \arg\max_{\Phi_k} \ell_k^n((x_k^n; \phi_k)$$

and

$$\hat{\theta}^n = \arg\max_{\Theta} \ell_c^n((x_1^n, \ldots, x_d^n); \hat{\phi}_1^n, \ldots, \hat{\phi}_d^n, \theta).$$

White (1994) proves that under some regularity conditions the multistage quasi-maximum likelihood estimator is asymptotically gaussian (see Theorem 6.11 in White 1994) and in particular

$$(\bar{B}_n^0)^{-1/2} A_n^0 \sqrt{n}(\hat{\eta}^n - \eta^0) \xrightarrow{d} N(0, I),$$

where $\eta^0 = (\phi_1^0, \ldots, \phi_d^0, \theta^0)$ is the true value of the parameter and $A_n^0 = \mathbb{E}[H_n^0]$, where H_n^0 is a block hessian matrix of the type

$$H_n^0 = \begin{pmatrix} \nabla_{\phi_1\phi_1}\ell_1^n(X_1^n; \phi_1^0) & 0 & \cdots & 0 \\ 0 & \nabla_{\phi_2\phi_2}\ell_2^n(X_2^n; \phi_2^0) & \cdots & 0 \\ \cdots & \cdots & \cdots & \cdots \\ \nabla_{\phi_1\theta}\ell_c^n((X_1^n, \ldots, X_d^n); \eta^0) & \nabla_{\phi_2\theta}\ell_c^n((X_1^n, \ldots, X_d^n); \eta^0) & \cdots & \nabla_{\theta\theta}\ell_c^n((X_1^n, \ldots, X_d^n); \eta^0). \end{pmatrix}$$

Moreover, $\bar{B}_n^0 = \mathbb{E}[B_n^0]$, where

$$B_n^0 = \begin{pmatrix} \frac{1}{n}\sum_{t=1}^n [s_{1t}^0 \cdot (s_{1t}^0)'] & \frac{1}{n}\sum_{t=1}^n [s_{1t}^0 \cdot (s_{2t}^0)'] & \cdots & \frac{1}{n}\sum_{t=1}^n [s_{1t}^0 \cdot (s_{ct}^0)'] \\ \frac{1}{n}\sum_{t=1}^n [s_{2t}^0 \cdot (s_{1t}^0)'] & \frac{1}{n}\sum_{t=1}^n [s_{2t}^0 \cdot (s_{2t}^0)'] & \cdots & \frac{1}{n}\sum_{t=1}^n [s_{2t}^0 \cdot (s_{ct}^0)'] \\ \cdots & \cdots & \cdots & \cdots \\ \frac{1}{n}\sum_{t=1}^n [s_{ct}^0 \cdot (s_{1t}^0)'] & \frac{1}{n}\sum_{t=1}^n [s_{ct}^0 \cdot (s_{2t}^0)'] & \cdots & \frac{1}{n}\sum_{t=1}^n [s_{ct}^0 \cdot (s_{ct}^0)'] \end{pmatrix},$$

where

$$s_{kt}^0 = \nabla_{\phi_k} \ln f_{kt}(x_{kt}; \phi_k^0), \quad k = 1, \ldots, d,$$

and

$$s_{ct}^0 = \nabla_\theta \ln c_t(F_{1t}(x_{1t}; \phi_1^0), F_{2t}(x_{2t}; \phi_2^0), F_{3t}(x_{3t}; \phi_3^0); \theta^0)$$

are the score functions (see White 1994 and Patton 2006a, b for more details). Gobbi (2014) proposes the problem of estimation of a *convolution-based time series* with the same technique proving that the same asymptotic results can be reached.

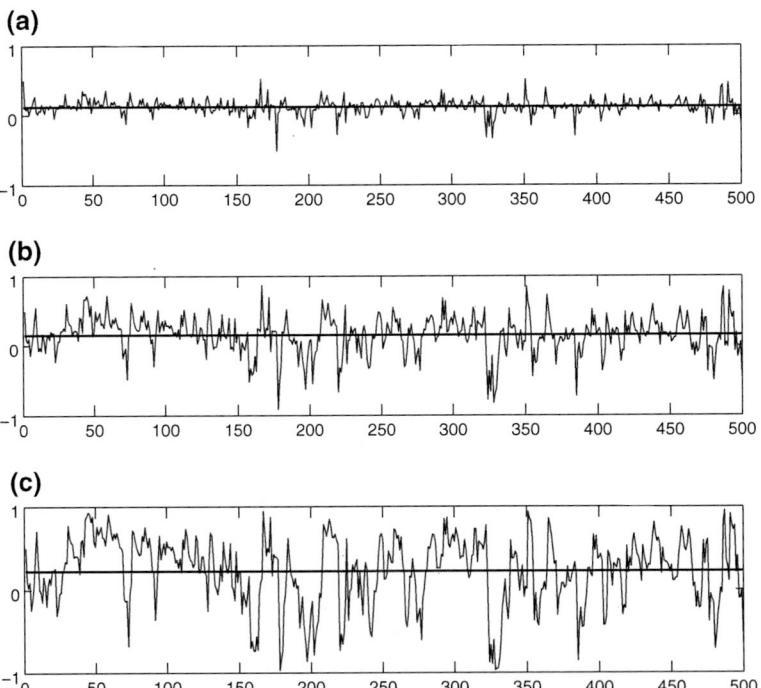

Fig. 2.5 Simulation of a time-varying correlation coefficient of a Gaussian copula with margins given by Garch(1,1) processes. The dynamics follows Eq. (2.5). **a** $\alpha = 0.2$, $\beta = 0.2$, $\gamma = 0.6$; **b** $\alpha = 0.2$, $\beta = 0.5$, $\gamma = 0.6$; **c** $\alpha = 0.2$, $\beta = 0.7$, $\gamma = 0.6$. The *solid line* is the average value

Example 2.6.1 In this example, we propose a simulation experiment based on Patton (2006a, b). In that paper the author assumes a bivariate copula-based time series where the conditional marginal distributions are both conditionally gaussian with the same Garch(1,1) specifications whose parameters are designed to reflect the highly persistence conditional volatility. In particular, the specification of the margins is as follows

$$X_t = \mu_X + h_t e_t$$
$$h_t^2 = \alpha_0 + \alpha_1 (X_{t-1} - \mu_X)^2 + \alpha_2 h_{t-1}^2$$
$$e_t | \mathcal{F}_{t-1} \overset{iid}{\sim} N(0, 1),$$

$$Y_t = \mu_Y + k_t q_t$$
$$k_t^2 = \beta_0 + \beta_1 (Y_{t-1} - \mu_Y)^2 + \beta_2 k_{t-1}^2$$
$$q_t | \mathcal{F}_{t-1} \overset{iid}{\sim} N(0, 1).$$

with $\mu_X = \mu_Y = 0.01$, $\alpha_0 = \beta_0 = 0.05$, $\alpha_1 = \beta_1 = 0.1$ and $\alpha_2 = \beta_2 = 0.85$. The time-dependent correlation coefficient is constructed according to two different dynamics, which are the following

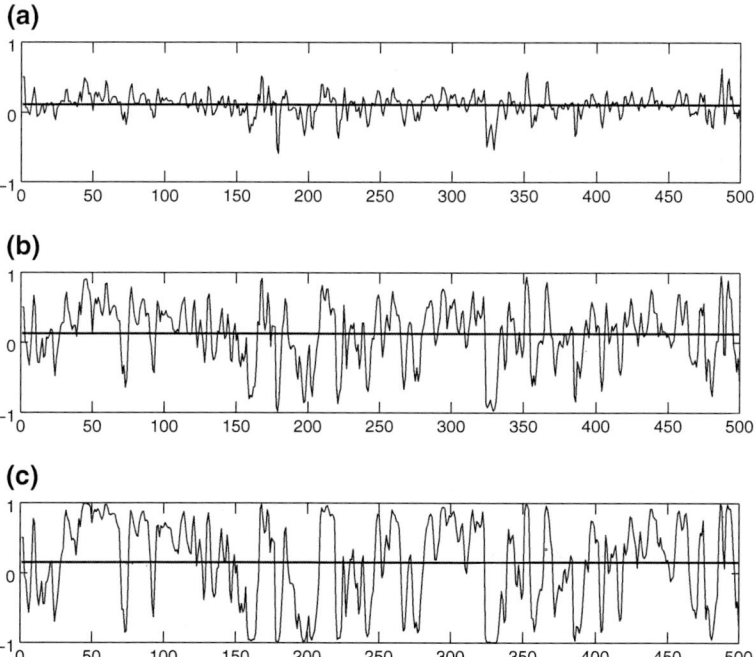

(a)

(b)

(c)

Fig. 2.6 Simulation of a time-varying correlation coefficient of a Gaussian copula with margins given by Garch(1,1) processes. The dynamics follows Eq. (2.6). **a** $\alpha = 0.2$, $\beta = 0.2$, $\gamma = 0.6$; **b** $\alpha = 0.2$, $\beta = 0.5$, $\gamma = 0.6$; **c** $\alpha = 0.2$, $\beta = 0.7$, $\gamma = 0.6$. The *solid line* is the average value

$$\rho_t = \Lambda\left(\alpha + \beta\rho_{t-1} + \delta\left[\Phi^{-1}(u_{t-1})\Phi^{-1}(v_{t-1})\right]\right), \tag{2.5}$$

$$\rho_t = \Lambda\left(\alpha + \beta\rho_{t-1} + \delta\left[\Phi^{-1}(u_{t-1})\Phi^{-1}(v_{t-1}) + \Phi^{-1}(u_{t-2})\Phi^{-1}(v_{t-2})\right]\right), \tag{2.6}$$

where $\Phi^{-1}(\cdot\,; \nu)$ is the inverse cdf of a standard gaussian distribution, $u_t = N(\mu, h_t)$ and $v_t = N(\mu, k_t)$. The function $\Lambda(\cdot)$ is a modified version of the logistic function $\Lambda(x) = (1 - e^{-x})/(1 + e^{-x})$ and it ensures that correlation coefficients ρ_t remain in the interval $(-1, 1)$ at all times. In practice, we set correlation coefficients at time t as a function of a constant, correlation coefficients at time $t - 1$ and some other variable which allows us to consider the time evolution in the dependence structure. The variables $\Phi^{-1}(u_{t-1})\Phi^{-1}(v_{t-1})$ and $\Phi^{-1}(u_{t-2})\Phi^{-1}(v_{t-2})$ deal with the influence of the two variables on their correlation coefficient. For our simulation we fix $\alpha = 0.2$, $\gamma = 0.6$ and $\beta = 0.2, 0.5, 0.7$ for both dynamics and we generate 500 sample points. Figures 2.5 and 2.6 display the simulated paths of the correlation coefficients.

References

Andersen, T. G., Bollerslev, T., Christoffersen, P., & Diebold, F. X. (2006). Volatility and correlation forecasting. *Handbook of economic forecasting* (Vol. 1, pp. 197–229). Oxford: Elsevier.

Durante, F., & Sempi, C. (2015). *Principles of copula theory.* Boca Raton: Chapman and Hall/CRC.

Genest, C., & Rivest, L.-P. (2001). On the multivariate probability integral transformation. *Statistics & Probability Letters, 53*, 391–399.

Gobbi, F. (2014). The conditional *C*-convolution model and the three stage quasi maximum likelihood estimator. *Journal of Statistics: Advances in Theory and Applications, 12*(1), 1–26.

Joe, H. (1990). Multivariate concordance. *Journal of Multivariate Analysis, 35*, 12–30.

Joe, H. (1997). *Multivariate models and dependence concepts.* London: Chapman & Hall.

Joe, H., & Xu, J. J. (1996). *The estimation method of inference functions for margins for multivariate models.* Technical report: Department of Statistics University of British Columbia. 166.

Kendall, M. G. (1938). A new measure of rank correlation. *Biometrika, 30*, 81–93.

Nelsen, R. B., Quesada-Molina, J. J., Rodríguez-Lallena, J. A., & Úbeda-Flores, M. (2003). Kendall distribution functions. *Statistics & Probability Letters, 65*, 263–268.

McNeil, A. J., & Nešlehová J. (2009). Multivariate Archimedean copulas, d-monotone functions and L1-norm symmetric distributions. *The Annals of Statistics, 37*(5B), 3059–3097.

Nelsen, R. N. (2006). *Introduction to copulas* (2nd ed.). Heidelberg: Springer.

Patton, A. J. (2001). Modelling time-varying exchange rate dependence using the conditional copula, UCSD Department of Economics, Working Paper.

Patton, A. J. (2006a). Modelling asymmetric exchange rate dependence. *International Economic Review, 47*(2), 527–556.

Patton, A. J. (2006b). Estimation of multivariate models for time series of possibly different lengths. *Journal of Applied Econometrics, 21*, 147–173.

Patton, A. J. (2012). A review of copula models for economic time series. *Journal of Multivariate Analysis, 110*, 4–18.

Rémillard, B. (2010). Goodness-of-fit tests for copulas of multivariate time series, SSRN Working Paper Series No. 1729982.

Ruymgaart, F. H., & van Zuijlen, M. C. A. (1978). Asymptotic normality of multivariate linear rank statistics in the non-i.i.d. case. *Annals of Statistics, 6*(3), 588–602.

Sklar, A. (1959). Fonctions de repartition à n dimensions et leurs marges. *Publications de l'Institut de statistique de l'Université de Paris, 8*, 229–231.

White, H. (1994). *Estimation, inference and specification analysis, econometric society monographs* (Vol. 22). Cambridge: Cambridge University Press.

Chapter 3
Copulas and Estimation
of Markov Processes

3.1 Copulas and Markov Processes: The DNO Approach

In this section, we briefly introduce a central result due to Darsow, Nguyen, and Olsen (see Darsow et al. 1992 for the original and complete result) that allows to characterize a Markov process through the dependence structure of the finite dimensional levels independently of their marginal distributions.

3.1.1 Markov Processes

Here we remind the definition of Markov process in discrete time and its main features.

Definition (*Markov Process*) Let $(\Omega, \mathcal{F}, (\mathcal{F}_t)_{t \in \mathbb{N}}, \mathbb{P})$ be a filtered probability space and $X = (X_t)_{t \in \mathbb{N}}$ be an adapted stochastic process. X is a Markov process if and only if

$$\mathbb{P}(X_t \leq x | X_{t-1}, X_{t-2}, \ldots, X_0) = \mathbb{P}(X_t \leq x | X_{t-1}). \tag{3.1}$$

Moreover, we call X a Markov process with respect to the filtration $(\mathcal{F}_t)_{t \in \mathbb{N}}$ if

$$\mathbb{P}(X_t \leq x | \mathcal{F}_s) = \mathbb{P}(X_t \leq x | X_s), \ s < t. \tag{3.2}$$

For every Borel set A, we set $P(s, x, t, A) = P(X_t \in A | X_s = x)$ the so-called transition probabilities.

It is a well-known fact (whose proof is a trivial check) that if X is a Markov process, than its transition probabilities satisfy the Chapman–Kolmogorov equation:

$$P(s, x, t, A) = \int_{-\infty}^{+\infty} P(u, \xi, t, A) P(s, x, u, d\xi), \ s < u < t.$$

© The Author(s) 2016
U. Cherubini et al., *Convolution Copula Econometrics*,
SpringerBriefs in Statistics, DOI 10.1007/978-3-319-48015-2_3

The first important result provided by Darsow, Nguyen, and Olsen consists in rewriting the Chapman–Kolmogorov equation in terms of copula functions. In order to do this, we need to introduce a suitable operation among bivariate copula functions. Nevertheless, Chapman–Kolmogorov equations are only a necessary condition for a process to be Markov. In order to state also a sufficient condition we need to introduce a further operator among multivariate copula functions. We devote next subsection to introduce these operators.

3.1.2 The ∗ and ⋆-Product Operators

Definition Let A and B be two bivariate copula functions. For $x, y \in [0, 1]$, the ∗ product operation is defined as follows:

$$(A * B)(x, y) = \int_0^1 D_2 A(x, t) D_1 B(t, y) dt, \tag{3.3}$$

where $D_i C(\cdot, \cdot), i = 1, 2$ denotes the partial derivative of copula $C(\cdot, \cdot)$ with respect to the the i-th argument.

We leave to the reader the check that $(A * B)(x, y)$ is again a copula. The only nontrivial point is the proof of the 2-increasing property. But

$$V_{(A*B)}([x_1, x_2], [y_1, y_2]) = \int_0^1 D_2[A(x_2, t) - A(x_1, t)]D_1[B(t, y_2) - B(t, y_1)]dt \tag{3.4}$$

and the result follows from the standard finding that both $D_2[A(x_2, t) - A(x_1, t)]$ and $D_1[B(t, y_2) - B(t, y_1)]$ are nonnegative (see Lemma 2.1.3 in Nelsen 2006).

One important property of the ∗ operator is that it is associative, that is

$$A * (B * C) = (A * B) * C \quad \text{for all } A, B, C \in \mathcal{C},$$

for whose proof we refer to the original Darsow et al. (1992) paper. Conversely, the following properties are left to the reader as exercises:

$$C * \Pi = \Pi * C = \Pi, \; C * M = M * C = C,$$

while

$$C * W(u, v) = u - C(1 - u, v), \; W * C(u, v) = v - C(1 - u, v)$$

and

$$W * W = M.$$

For other properties of the ∗-product, we also refer the reader to Cherubini et al. (2012).

Let us introduce another operator among copulas, that is a generalization of the one just introduced.

Definition Let A be an m-copula and B an n-copula. The $\star$-product operation is defined as follows:

$$(A \star B)(u_1, \ldots, u_{m+n-1}) =$$

$$= \int_0^{x_m} D_m A(u_1, \ldots, u_{m-1}, \xi) D_1 B(\xi, x_{m+1}, \ldots, x_{m+n-1}) d\xi,$$

where we remind that $D_i C$ denotes the partial derivative of copula C with respect to the the i-th argument.

By arguments similar to those used for the ∗ product it is easily verified that $A \star B$ is an $(n + m - 1)$-copula and satisfies the same properties listed above.

Notice that $A \ast B(x, y) = A \star B(x, 1, y)$ implies that the $\star$-product is indeed a generalization of the ∗-product.

3.1.3 The Darsow, Nguyen, and Olsen Theorem

We are now ready to introduce the main results of Darsow et al. (1992) paper; we refer the reader to the original paper for the detailed proofs.

The first one introduces a characterization of the Chapman–Kolmogorov equation in terms of the ∗-product among copula functions.

Theorem 3.1.1 *Let $(X_t)_{t \in \mathbb{N}}$ be a real stochastic process, and for each $s, t \in \mathbb{N}$ let C_{st} denote the copula of the random variables X_s and X_t. The following are equivalent:*

(i) *The transition probabilities $P(s, x, t, A) = P(X_t \in A | X_s = x)$ of the process satisfy the Chapman–Kolmogorov equations*

$$P(s, x, t, A) = \int_{-\infty}^{+\infty} P(u, \xi, t, A) P(s, x, u, d\xi),$$

for all Borel sets A, for all $s < u < t$ in $\mathbb{N}$ and for almost all $x \in \mathbb{R}$.

(ii) *For all $s < u < t$ in $\mathbb{N}$*

$$C_{st} = C_{su} \ast C_{ut}. \tag{3.5}$$

Next one is the main result, where the Markov property is characterized through the $\star$-product.

Theorem 3.1.2 *A real valued stochastic process* $(X_t)_{t \in \mathbb{N}}$ *is a Markov process if and only if for all positive integers n and for all* $t_1, \ldots, t_n \in \mathbb{N}$ *satisfying* $t_k < t_{k+1}, k = 1, \ldots, n - 1,$

$$C_{t_1 \ldots t_n} = C_{t_1 t_2} \star C_{t_2 t_3} \star \ldots \star C_{t_{n-1} t_n}, \tag{3.6}$$

where $C_{t_1 \ldots t_n}$ *is the copula of* $X_{t_1}, \ldots, X_{t_n}$ *and* $C_{t_k t_{k+1}}$ *is the copula of* X_{t_k} *and* $X_{t_{k+1}}$.

3.1.4 Building Technique of Markov Processes

The results presented in the previous subsection suggest a technique to construct Markov processes. The procedure consists in: (a) specifying a family of bivariate copulas satisfying the $*$-product closure condition (3.5); (b) specifying the marginal distributions.

More precisely, let $(C_{i,i+1})_{i \in \mathbb{N}}$ be a set of bivariate copula functions defining the dependence between the state of the process at time t_i and time t_{i+1}. Then, for $i \in \mathbb{N}$ and $k \geq 2$, consider the copulas

$$C_{i,i+k} = C_{i,i+1} * C_{i+1,i+2} \ldots * C_{i+k-1,i+k}.$$

Obviously, the class so defined is closed with respect to the $*$-product by definition.

In particular, if $C_{i,i+1} \equiv C$ for any i,

$$C_{i,i+k} = C * C * \ldots * C = C^k$$

is the k-fold $*$-product of C.

Since, the copulas $C_{i,i+k}$ are uniquely determined after having specified the starting family $C_{i,i+1}$, we are interested in analyzing the particular dependence structure that they induce on bivariate realizations on the process $(X_t)_{t \in \mathbb{N}}$ they are specifying. In particular we question if, assuming that the $C_{i,i+1}$ belong to the same family of copulas, so the induced $C_{i,k}$ do.

The Gaussian Family

The Gaussian family is closed with respect to the $*$-product. More precisely,

$$C_{i,i+1}(u, v) = \int_0^u \Phi \left(\frac{\Phi^{-1}(v) - \rho_{i,i+1} \Phi^{-1}(w)}{\sqrt{1 - \rho_{i,i+1}^2}} \right) dw, \ \rho_{i,i+1} \in [-1, 1]. \tag{3.7}$$

We have that

$$C_{i,i+k}(u, v) = \int_0^u \Phi\left(\frac{\Phi^{-1}(v) - \rho_{i,i+k}\Phi^{-1}(w)}{\sqrt{1 - \rho_{i,i+k}^2}}\right) dw$$

with

$$\rho_{i,i+k} = \prod_{j=0}^{k-1} \rho_{i+j,i+j+1}.$$

For a proof of this fact, see Cherubini et al. (2012), Sect. 3.2.3.

If $|\rho_{i,i+1}| \leq \rho \in (0, 1)$ (this happens, for example, in the stationary case $\rho_{i,i+1} \equiv \hat{\rho}$ for all i's), we get $\lim_{k \to +\infty} \rho_{i,i+k} = 0$, which is the dependence structure tends to independence.

The α-Stable Family

Let Z be a standard α-stable random variable, that is we assume that Z has a characteristic function of type

$$\phi_Z(t) = e^{-|t|^\alpha},$$

with $\alpha \in (0, 2]$. As shown Example 3.4.6 in Cherubini et al. (2012), if Φ_Z is the cumulative distribution function of Z, then

$$C_\rho(u, v) = \int_0^u \Phi_Z\left(\frac{\Phi_Z^{-1}(v) - \rho\Phi_Z^{-1}(w)}{(1 - \rho^\alpha)^{\frac{1}{\alpha}}}\right) dw, \quad \rho \in (0, 1)$$

is a copula function and for any $\rho_1, \rho_2 \in (0, 1)$

$$C_{\rho_1} * C_{\rho_2} = C_{\rho_1\rho_2}.$$

It follows that, if

$$C_{i,i+1}(u, v) = \int_0^u \Phi_Z\left(\frac{\Phi_Z^{-1}(v) - \rho_{i,i+1}\Phi_Z^{-1}(w)}{(1 - \rho_{i,i+1}^\alpha)^{\frac{1}{\alpha}}}\right) dw, \quad \rho \in (0, 1),$$

then

$$C_{i,i+k}(u, v) = \int_0^u \Phi\left(\frac{\Phi^{-1}(v) - \rho_{i,i+k}\Phi^{-1}(w)}{(1 - \rho_{i,i+k}^\alpha)^{\frac{1}{\alpha}}}\right) dw,$$

with

$$\rho_{i,i+k} = \prod_{j=0}^{k-1} \rho_{i+j,i+j+1}$$

As in the Gaussian case (that is a included in the present one when $\alpha = 2$), if $|\rho_{i,i+1}| \leq \rho \in (0, 1)$, we get $\lim\limits_{k \to +\infty} \rho_{i,i+k} = 0$, that is the dependence structure tends to independence.

The Farlie–Gumbel–Morgenstern Family

Now we assume

$$C_{i,i+1}(u, v) = uv + \theta_{i,i+1} uv(1 - u)(1 - v), \; \theta_{i,i+1} \in [-1, 1]. \tag{3.8}$$

We have that

$$C_{i,i+k}(u, v) = uv + \theta_{i,i+k} uv(1 - u)(1 - v),$$

with

$$\theta_{i,i+k} = 3^{-k+1} \prod_{j=0}^{k-1} \theta_{i+j,i+j+1}.$$

For a proof of this fact, see Cherubini et al. (2012), Sect. 3.2.3.

Since $|\theta_{i,i+k}| \leq \frac{1}{3^{k-1}}$, we have that $\lim\limits_{k \to +\infty} \theta_{i,i+k} = 0$, that is the dependence structure tends to independence.

Fréchet Dynamics

In this case

$$C_{\alpha,\beta} = \alpha W + \beta M + (1 - \alpha - \beta)\Pi, \; \alpha, \beta, 1 - \alpha - \beta \in [0, 1]$$

and

$$C_{\alpha,\beta} * C_{\hat{\alpha},\hat{\beta}} = C_{\alpha\hat{\beta}+\beta\hat{\alpha}, \alpha\hat{\alpha}+\beta\hat{\beta}}.$$

If

$$C_{i,i+1} = \alpha_{i,i+1} W + \beta_{i,i+1} M + (1 - \alpha_{i,i+1} - \beta_{i,i+1})\Pi,$$

then

$$C_{i,i+k} = \alpha_{i,i+k} W + \beta_{i,i+k} M + (1 - \alpha_{i,i+k} - \beta_{i,i+k})\Pi$$

with

$$\beta_{i,i+k} = \prod_{j=0}^{k-1} \alpha_{i+j,i+j+1} + \prod_{j=0}^{k-1} \beta_{i+j,i+j+1}$$

and $\alpha_{i,i+k}$ can be recursively obtained from

$$\alpha_{i,i+k} = \alpha_{i,i+k-1}\beta_{i+k-1,i+k} + \prod_{j=0}^{k-1} \alpha_{i+j,i+j+1} + \alpha_{i+k-1,i+k} \prod_{j=0}^{k-2} \beta_{i+j,i+j+1}. \tag{3.9}$$

If $\alpha_{i,i+1} \le \alpha \in (0,1)$ and $\beta_{i,i+1} \le \beta \in (0,1)$, we get $\lim\limits_{k\to+\infty} \beta_{i,i+k} = 0$ and $\lim\limits_{k\to+\infty} \alpha_{i,i+k} = 0$. This fact is obvious for $\beta_{i,i+k}$. As for $\alpha_{i,i+k}$, from (3.9) we get

$$\alpha_{i,i+k} \le \alpha_{i,i+k-1}\beta + \alpha^k + \alpha\beta^{k-1}.$$

Through recursion we get

$$\alpha_{i,i+k} \le \alpha_{i,i+1}\beta^{k-1} + \sum_{j=2}^{k}(\alpha^j + \alpha\beta^{j-1})\beta^{k-j}$$

from which

$$\alpha_{i,i+k} \le \alpha_{i,i+1}\beta^{k-1} + \frac{\alpha^2}{\beta-\alpha}\beta^{k-1} - \frac{\alpha^2}{\beta-\alpha}\alpha^{k-1} + \alpha(k-1)\beta^{k-1}$$

from which trivially follows that $\lim\limits_{k\to+\infty} \alpha_{i,i+k} = 0$.

3.2 Copula-Based Markov Processes: Estimation, Mixing Properties, and Long-Term Behavior

From an econometric point of view, copula-based Markov processes was introduced, discussed, and estimated in a seminal paper by Chen and Fan in 2006. Here the authors give a formal definition and establish some assumptions necessary to obtain asymptotic results relative to maximum likelihood estimators of the parameters. Their approach is semiparametric since whereas the copula function which describes the temporal dependence belongs to a parametric family the marginal distributions are given by empirical distribution functions.

A copula-based stationary Markov process is a first-order Markov process generated by a copula function which captures the temporal dependence of the process. More precisely, the process $X = (X_t)_{t\ge 0}$ is characterized by $\{G(\cdot, \eta), C(\cdot, \cdot; \theta)\}$, where G is the invariant distribution of X_t for all t with parameter (or vector of parameters) η and C is the copula function of (X_{t-1}, X_t) for all t and with parameter θ. In order to highlight the dynamic properties of a stationary copula-based Markov process, we present a simulation of trajectories with two types of temporal dependence structure given by two of the most used copula functions: Clayton and Frank.

Example 3.2.1 In this example, we show a simulation experiment to generate trajectories, denoted by $x = (x)_{t=1,...,n}$, from a strictly stationary copula-based Markov process characterized by the pair $\{G(\cdot, \eta), C(\cdot, \cdot; \theta)\}$ where G is a Student's t distribution with η degrees of freedom, whereas C is both a Clayton copula and a

(a)

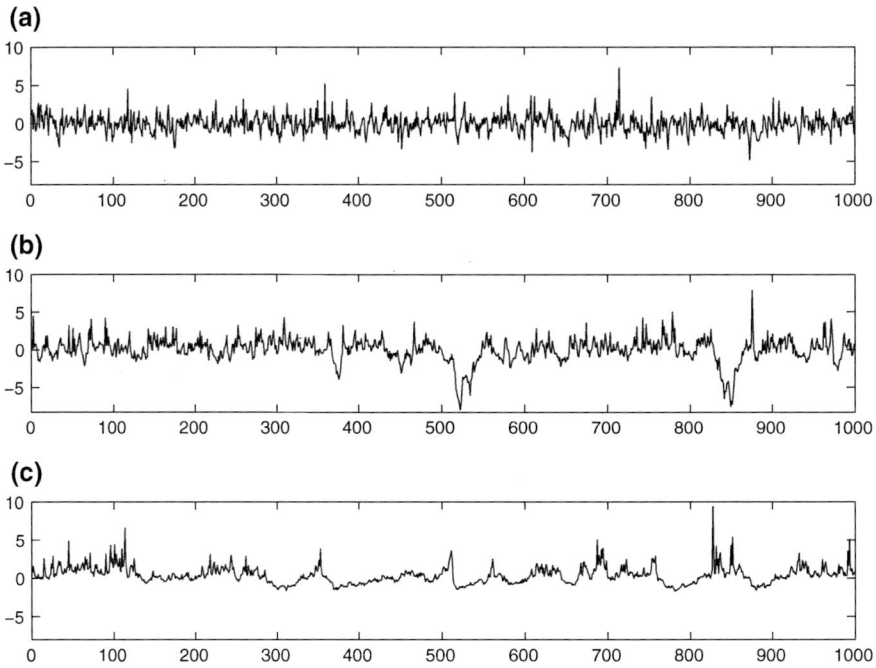

Fig. 3.1 Simulated trajectories from a copula-based Markov process when C is a Clayton copula and $G \sim t_{(5)}$. **a** $\tau = 0.2$ ($\theta = 0.5$); **b** $\tau = 0.5$ ($\theta = 2$); **c** $\tau = 0.8$ ($\theta = 8$)

Frank copula with parameter θ. The simulation procedure is the dynamic version of the conditional sampling method introduced in Sect. 2.4. First, we generate n iid $U(0, 1)$ samples $(\zeta_1, \ldots, \zeta_n)$ and then for each t we solve $u_t = h^{-1}(\zeta_t | u_{t-1})$, where $h(u_t | u_{t-1}) \equiv D_1 C(u_{t-1}, u_t)$. The desired sample is $(u_1, \ldots, u_n)$. To simulate the copula-based Markov time series $(x)_{t=1,\ldots,n}$ suffices to apply a generalized inverse of the stationary distribution G, that is $x_t = G^{-1}(u_t)$, $t = 1, \ldots, n$. We use a Student's t distribution with 5 degrees of freedom as invariant distribution. Moreover the parameters of the Clayton copula and of the Frank copula are chosen in such a way that they are consistent with three different levels of Kendall's τ: 0.2, 0.5 and 0.8. Figures 3.1 and 3.2 present time series plots. We notice the asymmetric dependence structure of the time series which becomes stronger as τ increases.

3.2.1 Semiparametric Estimation

This section introduces semiparametric estimators of a copula-based Markov model and establishes their asymptotic properties under easily verifiable condition. The main contribution is due to Chen and Fan (2006) (whereas an extension to multivariate

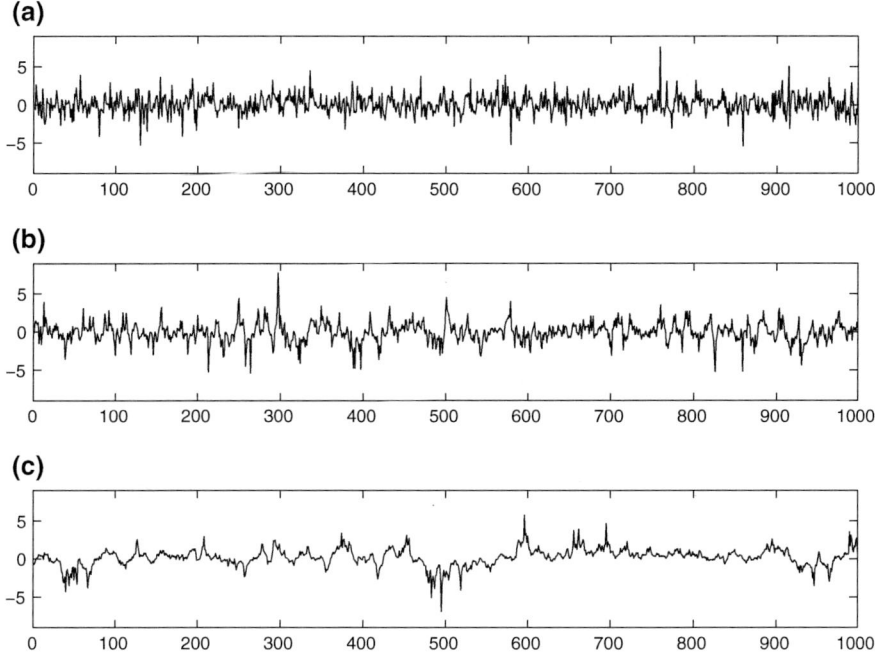

Fig. 3.2 Simulated trajectories from a copula-based Markov process when C is a Frank copula and $G \sim t_{(5)}$. **a** $\tau = 0.2$ ($\theta = 1.86$); **b** $\tau = 0.5$ ($\theta = 5.74$); **c** $\tau = 0.8$ ($\theta = 18.19$)

contexts can be find in Remillard et al. 2012). In that paper, the authors study a class of univariate copula-based semiparametric stationary Markov models in which copulas are parameterized but the invariant marginal distributions are left unspecified. In these models there are only two parameters: the copula dependence parameter θ and the invariant marginal distribution function $G(\cdot)$ which can be estimated by a nonparametric method, i.e., the empirical distribution function or the kernel smoothed estimator.

Let $X = (X_t)_{t \in \mathbb{Z}}$ be a stationary Markov process generated by $\{G^*(\cdot, \eta^*),$ $C^*(\cdot, \cdot; \theta^*)\}$, where G^* denotes the true invariant marginal distribution and C^* is the true copula function which describes the temporal dependence structure. Let $H^*(x, y)$ be the joint distribution function of X_{t-1} and X_t. We know that H^* completely determines the probabilistic structure of X. If C^* is the copula between X_{t-1} and X_t, we model a stationary Markov process by specifying the marginal distribution of X_t and the copula C^* instead of the joint distribution H^*, which will be $H^*(x_1, x_2) = C^*(G^*(x_1), G^*(x_2); \theta)$.

The conditional density of X_t given X_{t-1} is given by

$$h^*(X_t | X_{t-1}) = g^*(X_t)c(G^*(X_{t-1}), G^*(X_t); \theta^*), \tag{3.10}$$

where $c(\cdot, \cdot; \theta^*)$ is the copula density and $g^*(\cdot)$ is the density of the true invariant distribution $G^*(\cdot)$.

A semiparametric copula-based time series model is completely determined by (G^*, θ^*). The unknown invariant marginal distribution G^* can be estimated by $G_n(\cdot)$, the rescaled empirical distribution function defined as

$$G_n(x) = \frac{1}{n+1} \sum_{t=1}^{n} \mathbf{1}_{\{X_t \le x\}}. \tag{3.11}$$

Since the log-likelihood function is given by

$$\ell(\theta) = \frac{1}{n} \sum_{t=1}^{n} \log g^*(X_t) + \frac{1}{n} \sum_{t=1}^{n} \log c(G^*(X_{t-1}), G^*(X_t); \theta),$$

ignoring the first term and replacing G^* with G_n we obtain the semiparametric estimator $\tilde{\theta}$ of θ^*

$$\tilde{\theta} = \arg \max_{\theta} \tilde{L}(\theta), \quad \tilde{L}(\theta) = \frac{1}{n} \sum_{t=1}^{n} \log c(G_n(X_{t-1}), G_n(X_t); \theta).$$

The main difficult in establishing the asymptotic properties of the semiparametric estimator $\tilde{\theta}$ is that the score function and its derivatives could blow up to infinity near the boundaries. To overcome this difficulty, Chen and Fan (2006) first established convergence of $G_n(\cdot)$ in a weighted metric and then use it to establish the consistency and asymptotic normality of $\tilde{\theta}$. Let $\tilde{U}_n(z) = G_n(G^{*-1}(z)), z \in]0, 1[$ and let $w(\cdot)$ be a continuous function on $[0, 1]$ which is strictly positive on $]0, 1[$ symmetric at $z = 1/2$ and increasing on $]0, 1/2]$ and suppose that X_t is β-mixing (see Sect. 3.2.2), then (Chen and Fan 2006, Lemma 4.1)

$$\sup_{z \in [0,1]} \left| \frac{\tilde{U}_n(z) - z}{w(z)} \right| = o_{a.s}(1), \quad \sup_{y} \left| \frac{G_n(y) - G^*(y)}{w(G^*(y))} \right| = o_{a.s}(1),$$

provided $\beta_t = O(t^{-b})$ for some $b > 0$ and $\int_0^1 \frac{1}{w(z)} \log(1 + \frac{1}{w(z)}) dz < \infty$. Weight functions of the form $w(z) = [z(1 - z)]^{1-\xi}$ for all $z \in]0, 1[$ and some $\xi \in]0, 1[$ approach zero when z approaches 0 or 1. Hence such results allow us to handle unbounded score functions.

The maximum likelihood estimator $\tilde{\theta}$ is consistent and asymptotically Gaussian with rate of convergence $\sqrt{n}$ (Chen and Fan 2006, Propositions 4.2 and 4.3). In particular, let $l(u, v; \theta) = \log c(u, v; \theta)$ and denote $l_\theta(u, v; \theta) = \frac{\partial l(u,v;\theta)}{\partial \theta}$, $l_{\theta,\theta}(u, v; \theta) = \frac{\partial^2 l(u,v;\theta)}{\partial \theta \partial \theta}$, $l_{\theta,u}(u, v; \theta) = \frac{\partial^2 l(u,v;\theta)}{\partial \theta \partial u}$, and $l_{\theta,v}(u, v; \theta) = \frac{\partial^2 l(u,v;\theta)}{\partial \theta \partial v}$. Then, two conclusions hold

1. under some regularity conditions (C1–C2 and C4–C5 in Chen and Fan 2006, pp. 317–318) and if X_t is β-mixing with mixing decay rate $\beta_t = O(t^{-b})$ for some $b > 0$ we have $\|\tilde{\theta} - \theta^*\| = o_p(1)$;
2. under some regularity conditions (A1–A3 and A5–A6 in Chen and Fan 2006, pp. 318–319) and if X_t is β-mixing with mixing decay rate $\beta_t = O(t^{-b})$ for some $b > \gamma/(\gamma - 1)$ with $\gamma > 1$ we have $\sqrt{n}(\tilde{\theta} - \theta^*) \to N(0, B^{-1}\Sigma B)$, where $B = -\mathbb{E}[l_{\theta,\theta}(U_{t-1}, U_t; \theta^*)]$ and $\Sigma = \lim_{n \to \infty} Var(\sqrt{n}A_n^*)$, where

$$A_n^* = \frac{1}{n-1} \sum_{t=2}^{n} [l_\theta(U_{t-1}, U_t; \theta^*) + W_1(U_{t-1}) + W_2(U_t)],$$

$$W_1(U_{t-1}) = \int_0^1 \int_0^1 [\mathbf{1}\{U_{t-1} \leq u - u\}]l_{\theta,u}(u, v; \theta^*)c(u, v; \theta^*)dudv,$$

$$W_2(U_t) = \int_0^1 \int_0^1 [\mathbf{1}\{U_t \leq v - v\}]l_{\theta,v}(u, v; \theta^*)c(u, v; \theta^*)dudv,$$

where $U_t = G^*(X_t)$.

The assumption of β-mixing condition can be replaced with a so-called strong mixing condition but in this case the existence of finite higher order moments of the score function and its partial derivatives will be stronger than those for β-mixing processes. As many copula models have score functions blowing up at a fast rate, it is essential to maintain minimal requirements for the existence of finite higher order moments and this is a sufficient motivation to assume β-mixing instead of strong mixing.

An alternative estimation procedure for the copula parameter θ^* and the invariant distribution G^* is proposed by Chen et al. (2009). They call it sieve maximum likelihood estimation and it consists in approximating the unknown marginal density by a flexible parametric family of densities with increasing complexity (sieves) and then maximizing the joint likelihood w.r.t. the unknown copula parameter and the sieve parameters of the approximating marginal density. The sieve MLE of any smooth functionals of (θ^*, G^*) are root-n consistent, asymptotically normal, and efficient.

We write the true conditional density of X_t given X_{t-1} as in (3.10). Let

$$h(X_t|X_{t-1}; g, \theta) = g(X_t)c(G(X_{t-1}), G(X_t); \theta),$$

denote any candidate conditional density of X_t given X_{t-1}. Here, the parameters are g and θ. The log-likelihood associated to $h(X_t|X_{t-1}; \theta, g)$ is given by

$$l(\theta, g; X_t, X_{t-1}) = \log g(X_t) + \log c(G(X_{t-1}), G(X_t); \theta)$$

$$\equiv \log g(X_t) + \log c\left(\int \mathbf{1}\{y \leq X_{t-1}\}g(y)dy, \int \mathbf{1}\{y \leq X_t\}g(y)dy; \theta\right),$$

and the joint log-likelihood function of the data $\{X_t, t = 1, \ldots, n\}$ is

$$L_n(\theta, g) = \frac{1}{n} \sum_{t=2}^{n} l(\theta, g; X_t, X_{t-1}) + \frac{1}{n} \log g(X_1).$$

Let Θ be the parameter space for the copula parameter θ and let $\mathcal{G}$ be the parameter space for g. Then, the approximate sieve MLE $\hat{\gamma}_n = (\hat{\theta}_n, \hat{g}_n)$ is defined as

$$L_n(\hat{\theta}_n, \hat{g}_n) \geq \max_{\theta \in \Theta, g \in \mathcal{G}_n} L_n(\theta, g) - O_p(\delta_n^2),$$

where δ_n is a positive sequence such that $\delta_n = o(1)$ and $\mathcal{G}_n$ denote the sieve space (i.e., a sequence of finite dimensional parameter spaces that becomes dense as $n \to \infty$ in $\mathcal{G}$). The sieves used by Chen et al. (2009) for approximating the invariant density function are linear. Two examples are given by

$$\mathcal{G}_n = \left\{ g_{K_n} \in \mathcal{G} : g_{K_n} = \left(\sum_{k=1}^{K_n} a_k A_k(y) \right)^2, \int g_{K_n}(y) dy = 1 \right\}, \quad K_n \to \infty, \frac{K_n}{n} \to 0,$$

to approximate a square root density, and

$$\mathcal{G}_n = \left\{ g_{K_n} \in \mathcal{G} : g_{K_n} = \exp\left(\sum_{k=1}^{K_n} a_k A_k(y) \right), \int g_{K_n}(y) dy = 1 \right\}, \quad K_n \to \infty, \frac{K_n}{n} \to 0,$$

to approximate a log-density, where $\{A_k : k \geq 1\}$ consists of known basis functions and $\{a_k : k \geq 1\}$ is the collection of unknown sieve coefficients.

We conclude this section with the two main Theorems concerning consistency and asymptotic normality.

Theorem 3.2.1 (Chen et al. 2009) *Under some regularity conditions (Assumptions 3.1 and 3.2 in Chen et al. 2009), if $K_n \to \infty$ and $\frac{K_n}{n} \to 0$ we have $\|\hat{\gamma}_n - \gamma\| = o_p(1)$.*

Let $\rho : \Theta \times \mathcal{G} \to \mathbb{R}$ be a smooth functional and $\rho(\hat{\gamma}_n)$ be the plug-in sieve MLE of $\rho(\gamma)$.

Theorem 3.2.2 (Chen et al. 2009) *Under some regularity conditions (Assumptions 4.1–4.7 in Chen et al. 2009), if X_t is strictly stationary β-mixing we have $\sqrt{n}(\rho(\hat{\gamma}_n) - \rho(\gamma)) \to_d N(0, \|\frac{\partial \rho(\gamma^*)}{\partial \gamma}\|^2)$.*

3.2.2 Mixing Properties

As discussed in the previous section consistency and asymptotic normality of estimators of copula functions and their parameters are obtained under assumptions of weak dependence in the time series considered. Among other results, Beare (2010), Chen et al. (2009), and Ibragimov and Lentzas (2009) provide a study of persistence properties of of stationary copula-based Markov processes. For a wide review on mixing conditions see Bradley (2007). The most important dependence property of copula-based Markov processes is the β-mixing. In this subsection, we briefly discuss this notion.

Definition The β-mixing coefficients $\{\beta_k : k \in \mathbb{N}\}$ corresponding to the sequence of random variables $\{X_t\}$ are given by

$$\beta_k = \frac{1}{2} \sup_{m \in \mathbb{Z}} \sup_{\{A_i\},\{B_j\}} \sum_{i=1}^{I} \sum_{j=1}^{J} |\mathbb{P}(A_i \cap B_j) - \mathbb{P}(A_i)\mathbb{P}(B_j)|,$$

where the second supremum is taken over all finite partitions $\{A_1, \ldots, A_I\}$ and $\{B_1, \ldots, B_J\}$ of Ω such that $A_i \in \mathcal{F}_{-\infty}^{m}$ for each i and $B_j \in \mathcal{F}_{m+k}^{\infty}$ for each j.

The ρ-mixing coefficients $\{\rho_k : k \in \mathbb{N}\}$ are given by

$$\rho_k = \sup_{m \in \mathbb{Z}} \sup_{f,g} |Corr(f, g)|,$$

where the second supremum is taken over all square integrable random variables f and g measurable w.r.t. $\mathcal{F}_{-\infty}^{m}$ and $\mathcal{F}_{m+k}^{\infty}$ respectively, with positive and finite variance and where $Corr(f, g)$ denotes the correlation between f and g.

The α-mixing coefficients $\{\alpha_k : k \in \mathbb{N}\}$ are given by

$$\alpha_k = \sup_{m \in \mathbb{Z}} \sup_{A \in \mathcal{F}_{-\infty}^{m}, B \in \mathcal{F}_{m+k}^{\infty}} |\mathbb{P}(A \cap B) - \mathbb{P}(A)\mathbb{P}(B)|.$$

The definition of β-mixing coefficients is taken from Beare (2010) and an equivalent form was originally stated by Volkonskii and Rozanov (1959). The second definition of ρ-mixing appeared for the first time in Kolmogorov and Rozanov (1960), whereas the α-mixing coefficients are commonly attributed to Rosenblatt (1956). The relationships among those dependence conditions were investigated in recent papers, see among the others Chen and Fan (2006), Chen et al. (2009) and Beare (2010). α-mixing is a weaker dependence condition than both β- and ρ-mixing

$$\beta_k \to 0 \quad or \quad \rho_k \to 0 \quad \Rightarrow \quad \alpha_k \to 0, \quad k \to \infty.$$

This is guaranteed by the fact that $2\alpha_k \leq \beta_k$ and $4\alpha_k \leq \rho_k$ (Bradley 2007). On the other hand, neither of the β- and ρ-mixing conditions is implied by the other: $\beta_k \to 0 \nRightarrow \rho_k \to 0$ and $\rho_k \to 0 \nRightarrow \beta_k \to 0$.

The link between copula-based Markov processes and mixing properties was profitably investigated by Beare (2010). In particular, the author proves sufficient conditions for a geometric rate of mixing in this class of models. Geometric

β-mixing is established under a rather strong condition that rules out asymmetry and tail dependence in the copula function.

For the next theorem is necessary to introduce the concept of maximal correlation. The maximal correlation ρ_C of the copula C is defined as

$$\rho_C = \sup_{f,g} \left| \int_0^1 \int_0^1 f(u)g(v)C(du, dv) \right|,$$

where the supremum is taken over all $f, g \in \mathbb{L}^2([0, 1])$ such that $\int_0^1 f(u)du = \int_0^1 g(u)du = 0$ and $\int_0^1 f^2(u)du = \int_0^1 g^2(u)du = 1$.

Theorem 3.2.3 (Beare 2010) *If the copula C is symmetric and absolutely continuous with square integrable density c and moreover if $\rho_C < 1$ then there exists $A < \infty$ and $\gamma > 0$ such that $\beta_k \leq Ae^{-\gamma k}$ for all k.*

The condition $\rho_C < 1$ is satisfied for all absolutely continuous copulas with square integrable density such that c is positive a.e. on $[0, 1]$ as stated in Beare (2010). Commonly used parametric copula functions satisfying this condition include the FGM, Frank, and Gaussian copulas. Copulas exhibiting upper or lower tail dependence will not admit, in general, square integrable density and the theorem cannot directly be applied. However, Chen et al. (2009) show that Markov processes generated via tail dependent copulas such as Clayton, Gumbel, and Student's t are geometric β-mixing.

As for geometric ρ-mixing Beare (2010) obtained a weaker condition that permits both asymmetry and tail dependence as stated in the next theorem.

Theorem 3.2.4 (Beare 2010) *If $\rho_C < 1$, then there exist $A < \infty$ and $\gamma > 0$ such that $\rho_k \leq Ae^{-\gamma k}$ for all k.*

A simple condition to verify that $\rho_C < 1$ for a specific copula function is provided (Beare 2010): the density of the absolutely part of C must be bounded away from zero on a set of measure one. Two examples of copulas satisfying this condition are given by the t-copula and Marshall-Olkin copula.

3.2.3 Long-Term Behavior

The analysis of long-term properties of a copula-based Markov time series is very significant in econometrics. Stationary Markov processes are regarded as examples of short-memory processes because the value of such process at a given time depends only on the value at the previous time and the analysis of how the properties of the copula function which generates the process affect the long-term behavior is very interesting. This class of models deals with nonlinear dependencies because as many authors observed (see, among others, Embrechts et al. 2002; McNeil et al. 2005; Granger 2003) the autocorrelation function is problematic in many settings, including the departure from Gaussianity and elliptic distributions that is common in financial market data and they are not appropriate in describing persistence in unconditional distributions.

The definition of long memory process $(X_t)_t$ is given by Granger (2003). He uses the (copula-linked) Hellinger measure of dependence $H(t, h) = H_{X_t, X_{t+h}}$ between r.vs. X_t and X_{t+h}

$$H(t, h) = 1 - \int_{-\infty}^{+\infty} \int_{-\infty}^{+\infty} f_{X_t, X_{t+h}}^{1/2}(x, y) f_{X_t}^{1/2}(x) f_{X_{t+h}}^{1/2}(y) dx dy$$

where $f_{X_t, X_{t+h}}$ is the joint density of (X_t, X_{t+h}) and f_{X_t}, $f_{X_{t+h}}$ are the corresponding marginal densities.

Definition A process $(X_t)_t$ is a long memory process if for some constant $A > 0$

$$H(t, h) \sim Ah^{-p}, \quad h \to \infty,$$

where $p > 0$; it is a short memory process if

$$H(t, h) = O(e^{-Ah}), \quad h \to \infty.$$

Ibragimov and Lentzas (2009) study the long memory properties of copula-based stationary Markov processes using a number of dependence measures between X_t and X_{t+h} expressed in terms of their copula $C_{t,t+h}(u, v)$ which is obtained by iterating the product copula $C_{t,t+h} = C_{t,t+1} * C_{t,t+h-1} = C * C_{t,t+h-1}, h = 2, \ldots,$ where C is the invariant copula between two consecutive r.v.s of the process X_t and X_{t+1} for any t. Since the copula-based Markov process is stationary every measure of temporal dependence is a function of h only. For a copula-based Markov process the authors give the definition of Hellinger measure of dependence in terms of the copula between X_t and X_{t+h} which is

$$H(h) = H_{X_t, X_{t+h}} = \frac{1}{2} \int_0^1 \int_0^1 [c_{t,t+h}^{1/2} - 1]^2 (u, v) du dv - 1$$

where $c_{t,t+h}$ is the copula density associated to (X_t, X_{t+h}). The main conclusion of Ibragimov and Lentzas (2009) is that there exist Clayton copula-based stationary Markov processes that exhibit long memory on the level of copulas. In contrast, Gaussian, and EFGM copulas produce short-memory.

3.3 k-th Order Markov Processes

The definition of Markov process can be extended to k-th order Markov process. In this Section, we will present the main results of the paper of Ibragimov and Lentzas (2009), where the k-th order Markov processes are studied and characterized in terms of the associated finite-dimensional copula functions

Definition (*k-order Markov Process*) Let $(\Omega, \mathcal{F}, (\mathcal{F}_t)_{t\in\mathbb{N}}, \mathbb{P})$ be a filtered probability space and $X = (X_t)_{t\in\mathbb{N}}$ be an adapted stochastic process. X is a Markov process of order $k \geq 1$ if and only if

$$\mathbb{P}(X_t \leq x | X_{t-1}, X_{t-2}, \ldots, X_0) = \mathbb{P}(X_t \leq x | X_{t-1}, \ldots, X_{t-k}) \tag{3.12}$$

In what follows, we assume that all copulas considered are absolutely continuous and the processes under study have continuous univariate cumulative distribution functions.

In order to state the main result, we need to introduce some concepts and notations.

Let $m, n \geq k \geq 1$ and A and B be, respectively, m- and n-dimensional copulas. Set

$$A_{1,\ldots,m|m-k+1,\ldots,m}(u_1, \ldots, u_{m-k}, \xi_1, \ldots, \xi_k) = \frac{\frac{\partial^k A(u_1,\ldots,u_{m-k},\xi_1,\ldots,\xi_k)}{\partial\xi_1,\ldots,\partial\xi_k}}{\frac{\partial^k A(1,\ldots,1,\xi_1,\ldots,\xi_k)}{\partial\xi_1,\ldots,\partial\xi_k}}, \quad \text{and}$$

$$B_{1,\ldots,n|1,\ldots,k}(\xi_1, \ldots, \xi_k, u_{m+1}, \ldots, u_{m+n-k}) = \frac{\frac{\partial^k B(\xi_1,\ldots,\xi_k,u_{m+1},\ldots,u_{m+n-k})}{\partial\xi_1,\ldots,\partial\xi_k}}{\frac{\partial^k B(\xi_1,\ldots,\xi_k,1,\ldots,1)}{\partial\xi_1,\ldots,\partial\xi_k}}.$$

Recall that, if $(Y_1, \ldots, Y_n)$ is a random vector with associated copula function $C(u_1, \ldots, u_n)$ and margins F_i for $i = 1, \ldots, n$,

$$\mathbb{P}(Y_1 \leq y_1, \ldots, Y_{n-k} \leq y_{n-k} | Y_{n-k+1} = y_{n-k+1}, \ldots, Y_n = y_n) =$$

$$= \frac{\frac{\partial^k C(F_1(y_1),\ldots,F_{n-k}(y_{n-k}),F_{n-k+1}(y_{n-k+1}),\ldots,F_n(y_n))}{\partial u_{n-k+1}\cdots\partial u_n}}{\frac{\partial^k C(1,1,\ldots,1,F_{n-k+1}(y_{n-k+1}),\ldots,F_n(y_n))}{\partial u_{n-k+1}\cdots\partial u_n}},$$

that is

$$\mathbb{P}(Y_1 \leq y_1, \ldots, Y_{n-k} \leq y_{n-k} | Y_{n-k+1} = y_{n-k+1}, \ldots, Y_n = y_n) =$$

$$= C_{1,\ldots,n|n-k+1,\ldots,n}(F_1(y_1), \ldots, F_n(y_n))$$

and, similarly,

$$\mathbb{P}(Y_{k+1} \leq y_{k+1}, \ldots, Y_n \leq y_n | Y_1 = y_1, \ldots, Y_k = y_k) = C_{1,\ldots,n|1,\ldots,k}(F_1(y_1), \ldots, F_n(y_n)).$$

Definition If $A(1, \ldots, 1, \xi_1, \ldots, \xi_k) = B(\xi_1, \ldots, \xi_k, 1, \ldots, 1) = C(\xi_1, \ldots, \xi_k)$, where C is k-variate copula, we define the $\star^k$-product of the copulas A and B as the copula $D = A \star^k B : [0, 1]^{m+n-k} \to [0, 1]$ given by

$$D(u_1, \ldots, u_{m+n-k}) = \int_0^{u_{m-k+1}} \cdots \int_0^{u_m} A_{1,\ldots,m|m-k+1,\ldots,m}(u_1, \ldots, u_{m-k}, \xi_1, \ldots, \xi_k) \cdot$$
$$\cdot B_{1,\ldots,n|1,\ldots,k}(\xi_1, \ldots, \xi_k, u_{m+1}, \ldots, u_{m+n-k}) dC(\xi_1, \ldots, \xi_k).$$

The $\star^k$-operator is a generalization of the $\star$-operator considered in Darsow et al. (1992); in fact, the Darsow et al. (1992) $\star$-operator is the $\star^k$ operator when $k = 1$. In particular, the $\star^k$ operator satisfies the same properties as $\star$. An analogous of Theorem 3.1.2 holds for k-th order Markov processes.

Theorem 3.3.1 *A real-valued stochastic process* $(X_t)_{t \in \mathbb{N}}$ *is a Markov process of order k, $k \geq 1$, if and only if for all $t_i \in \mathbb{N}$, $i = 1, \ldots, n$, $n \geq k + 1$, such that $t_1 < \cdots < t_n$,*

$$C_{t_1,\ldots,t_n} = C_{t_1,\ldots,t_{k+1}} \star^k C_{t_2,\ldots,t_{k+2}} \star^k \cdots \star^k C_{t_{n-k},\ldots,t_n}.$$

We refer the interested reader to Ibragimov and Lentzas (2009) for the proof.

References

Beare, B. (2010). Copulas and temporal dependence. *Econometrica*, *78*(1), 395–410.

Bradley, R. C. (2007). *Introduction to strong mixing conditions*. Heber City: Kendrick Press.

Chen, X., & Fan, Y. (2006). Estimation of copula-based semiparametric time series models. *Journal of Econometrics*, *130*, 307–335.

Chen, X., Wu, W. B., & Yi, Y. (2009). Efficient estimation of copula-based semiparametric Markov models. *Annals of Statistics*, *37*(6B), 4214–4253.

Cherubini, U., Gobbi, F., Mulinacci, S., & Romagnoli, S. (2012). *Dynamic copula methods in finance*. New York: Wiley.

Darsow, W. F., Nguyen, B., & Olsen, E. T. (1992). Copulas and Markov processes. *Illinois Journal of Mathematics*, *36*, 600–642.

Embrechts, P., McNeil, A., & Straumann, D. (2002). Correlation and dependence in risk management: properties and pitfalls. In M. A. H. Dempster (Ed.), *Risk management: Value at risk and beyond* (pp. 176–223). Cambridge: Cambridge University Press.

Granger, C. J. W. (2003). Time series concepts for conditional distributions. *Oxford Bulletin for Economics and Statistics*, *65*, 689–701.

Ibragimov, R., & Lentzas, G. (2009). Copulas and long memory, Harvard Institute of Economic Research Discussion Paper No. 2160. Revised November 2009.

Kolmogorov, A. N., & Rozanov, Y. A. (1960). On the strong mixing conditions for stationary Gaussian sequences. *Theory of probability and its Applications*, *5*, 204–207.

McNeil, A. J., Frey, R., & Embrechts, P. (2005). *Quantitative risk management. Concepts, techniques and tools*. Princeton series in finance. Princeton, NJ: Princeton University Press.

Nelsen, R. N. (2006). *Introduction to copulas* (2nd ed.). Heidelberg: Springer.

Remillard, B., Papageorgiou, N., & Soustra, F. (2012). Copula-based semiparametric models for multivariate time series. *Journal of Multivariate Analysis*, *110*, 30–42.

Rosenblatt, M. (1956). A central limit theorem and a strong mixing condition. *Proceedings of the National Academy of Sciences of the United States of America*, *42*, 43–47.

Volkonskii, A., & Rozanov, Yu. A. (1959). Some limit theorems for random functions. *Theory of Probability and Its Applications*, *4*(2), 178–197.

Chapter 4
Convolution-Based Processes

4.1 The C-Convolution and Convolution-Based Copulas

In what follows, we consider a random vector (X, Y) and we study the distribution of $X + Y$ and the copula associated to the random vector $(X, X + Y)$. Since this represents the basic concept of the book, we include proofs, even if they are also presented in Cherubini et al. (2012) (see also Cherubini et al. 2011).

In the sequel, we assume continuous and strictly increasing marginal cumulative distribution functions.

Proposition 4.1.1 *Let X and Y be two real-valued random variables on the same probability space $(\Omega, \Im, \mathbb{P})$ with corresponding copula $C_{X,Y}$ and marginals F_X and F_Y. Then,*

$$C_{X,X+Y}(u, v) = \int_0^u D_1 C_{X,Y}\left(w, F_Y(F_{X+Y}^{-1}(v) - F_X^{-1}(w))\right) dw \qquad (4.1)$$

and

$$F_{X+Y}(t) = \int_0^1 D_1 C_{X,Y}\left(w, F_Y(t - F_X^{-1}(w))\right) dw. \qquad (4.2)$$

Proof

$$F_{X,X+Y}(s, t) = \mathbb{P}\left(X \leq s, X + Y \leq t\right) =$$

$$= \int_{-\infty}^s \mathbb{P}\left(X + Y \leq t | X = x\right) dF_X(x) =$$

$$= \int_{-\infty}^s \mathbb{P}\left(Y \leq t - x | X = x\right) dF_X(x) =$$

$$= \int_{-\infty}^s D_1 C_{X,Y}\left(F_X(x), F_Y(t - x)\right) dF_X(x) =$$

$$= \int_0^{F_X(s)} D_1 C_{X,Y}\left(w, F_Y(t - F_X^{-1}(w))\right) dw,$$

© The Author(s) 2016
U. Cherubini et al., *Convolution Copula Econometrics*,
SpringerBriefs in Statistics, DOI 10.1007/978-3-319-48015-2_4

where we made the substitution $w = F_X(x) \in (0, 1)$.

Then, the copula function linking X and $X + Y$ is

$$C_{X,X+Y}(u, v) = F_{X,X+Y}\left(F_X^{-1}(u), F_{X+Y}^{-1}(v)\right) =$$
$$= \int_0^u D_1 C_{X,Y}\left(w, F_Y(F_{X+Y}^{-1}(v) - F_X^{-1}(w))\right) dw.$$

Moreover,

$$F_{X+Y}(t) = \lim_{s \to +\infty} F_{X,X+Y}(s, t) = \int_0^1 D_1 C_{X,Y}\left(w, F_Y(t - F_X^{-1}(w))\right) dw.$$

□

The above result allows to introduce the notion of *C-convolution*.

Definition Let F, H be two cumulative distribution functions and C a copula function. We define the **C-convolution** of H and F the cumulative distribution function

$$H \overset{C}{*} F(t) = \int_0^1 D_1 C\left(w, F(t - H^{-1}(w))\right) dw.$$

Notice that we recover the standard notion of *convolution* if $C \equiv \Pi$.

An interesting property of the C-convolution, it is that it is closed with respect to mixtures of copula functions. In fact, if $C(u, v) = \lambda A(u, v) + (1 - \lambda)B(u, v)$ for $\lambda \in [0, 1]$ and A and B copula functions, then, for all c, H and F,

$$H \overset{C}{*} F = H \overset{\lambda A+(1-\lambda)B}{*} F = \lambda H \overset{A}{*} F + (1 - \lambda) H \overset{B}{*} F.$$

Another important consequence of the above Proposition is the introduction of the *Convolution-based copula function*.

Definition Let F and H be two cumulative distribution functions and $C(\cdot, \cdot)$ a copula function. Then

$$\hat{C}(u, v) = \int_0^u D_1 C\left(w, F\left((H \overset{C}{*} F)^{-1}(v) - H^{-1}(w)\right)\right) dw.$$

is called the *Convolution-based copula function*.

If F and C are absolutely continuous and f and c are their corresponding densities,

$$g^C(t) = \int_0^1 c\left(w, F(t - H^{-1}(w))\right) f\left(t - H^{-1}(w)\right) dw$$

is the density of $H \overset{C}{*} F$ provided that the above integral exists. Moreover, if g^C is positive,

$$\hat{c}(u, v) = c\left(u, F\left((H \overset{C}{*} F)^{-1}(v) - H^{-1}(u)\right)\right) \frac{f\left(u, F\left((H \overset{C}{*} F)^{-1}(v) - H^{-1}(u)\right)\right)}{g^C\left((H \overset{C}{*} F)^{-1}(v)\right)}$$

is the density of $\hat{C}$.

Example 4.1.1 **Archimedean Copulas**

1. **The Clayton Copula**

$$C(u, v) = \max\left([u^{-\theta} + v^{-\theta} - 1]^{-1/\theta}, 0\right) \quad \theta \in [-1, \infty) \setminus \{0\}.$$

Since $D_1 C(u, v) = \max\left([u^{-\theta} + v^{-\theta} - 1]^{-1/\theta-1} u^{-1-\theta}, 0\right)$

$$H \overset{C}{*} F(t) = \int_0^1 \max\left([w^{-\theta} + F(t - H^{-1}(w))^{-\theta} - 1]^{-1/\theta-1} w^{-1-\theta}, 0\right) dw$$

and

$$\hat{C}(u, v) = \int_0^u \max\left([w^{-\theta} + F((H \overset{C}{*} F)^{-1}(v) - H^{-1}(w))^{-\theta} - 1]^{-1/\theta-1} w^{-1-\theta}, 0\right) dw.$$

2. **The Gumbel Copula**

$$C(u, v) = uv \exp(-\theta \ln u \ln v), \quad \theta \in (0, 1].$$

Since $D_1 C(u, v) = (1 - \theta)v \exp(-\theta \ln u \ln v)$

$$H \overset{C}{*} F(t) = (1 - \theta) \int_0^1 F(t - H^{-1}(w)) \exp(-\theta \ln w \ln F(t - H^{-1}(w))) dw$$

and

$$\hat{C}(u, v) = (1 - \theta) \int_0^u F((H \overset{C}{*} F)^{-1}(v) - H^{-1}(w))$$

$$\exp(-\theta \ln w \ln F((H \overset{C}{*} F)^{-1}(v) - H^{-1}(w))) dw.$$

3. **The Frank Copula**

$$C(u, v) = -\frac{1}{\theta} \ln \left(1 + \frac{(e^{-\theta u} - 1)(e^{-\theta v} - 1)}{e^{-\theta} - 1} \right), \quad \theta \in \mathbb{R} \setminus \{0\}.$$

Since $D_1 C(u, v) = \frac{e^{-\theta u}(e^{-\theta v} - 1)}{e^{-\theta} - 1 + (e^{-\theta u} - 1)(e^{-\theta v} - 1)}$

$$H \overset{C}{*} F(t) = \int_0^1 \frac{e^{-\theta w}(e^{-\theta F(t - H^{-1}(w))} - 1)}{e^{-\theta} - 1 + (e^{-\theta w} - 1)(e^{-\theta F(t - H^{-1}(w))} - 1)} dw$$

and

$$\hat{C}(u, v) = \int_0^u \frac{e^{-\theta w}(e^{-\theta F((H \overset{C}{*} F)^{-1}(v) - H^{-1}(w))} - 1)}{e^{-\theta} - 1 + (e^{-\theta w} - 1)(e^{-\theta F((H \overset{C}{*} F)^{-1}(v) - H^{-1}(w))} - 1)} dw.$$

Clearly, in none of the above examples the C-convolution-based copula can be explicitly computed and, in particular, none is of the same family of the starting copula C. Hence, both the C-convolution and the C-convolution-based copula, can be only numerically estimated or their induced distribution be simulated. So, a discrete approximation by numerical integration is necessary. Among several methods of numerical integration, the simplest way is the following: given a sufficiently dense partition $\{w_0, \ldots, w_n\}$ of the interval $[0, 1]$ such that $0 \leq w_0 \leq \cdots \leq w_n \leq 1$, a simple approximation of g^C at the point x is given by

$$g^C(x) \simeq \sum_{i=1}^n c(w_{i-1}, F(t - H^{-1}(w_{i-1}))) f(x - H^{-1}(w_{i-1}))(w_i - w_{i-1}).$$

The approximation improves as n increases. Figure 4.1 displays the density of the C-convolution in the case where C is Frank or Clayton and the marginal distributions are standard normal and compares the shape of the C-convolution when the level of dependence, measured by Kendall's τ coefficient, varies in a range from 0.25 to 0.75. As expected, the Frank copula affects the tails of the distribution of the sum, whereas the Clayton copula affects the skewness. Figure 4.2 depicts the same C-convolution density when the second marginal distribution, F, is a Student's t with 3 degrees of freedom. The copula function in this case is the Frank copula with two levels of dependence (0.25 and 0.75).

4.1.1   Closure of C-Convolution

In this section we analyze those families of copulas in which, at least under suitable and very restrictive assumptions, the C-convolution operation is closed.

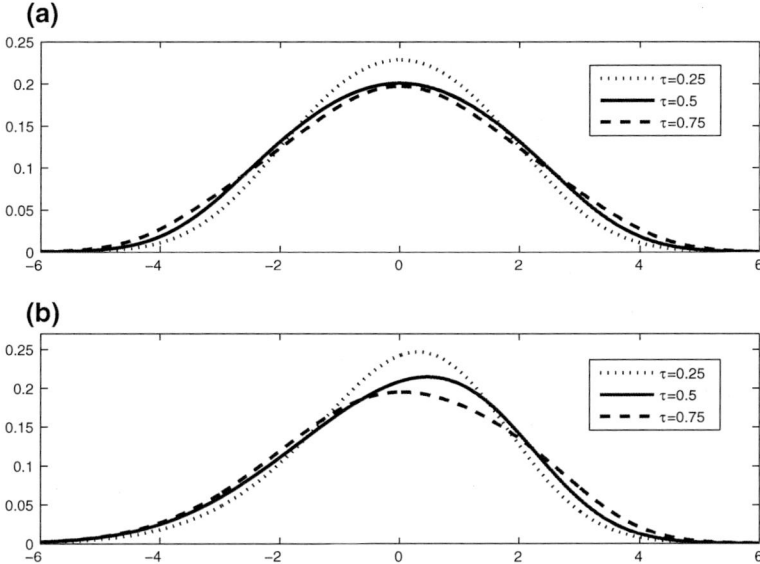

Fig. 4.1 Numerical approximation of the density of the C-convolution, g^C, when the marginal distribution F and H are standard Gaussian and the copula C is: **a** the Frank copula, **b** the Clayton copula

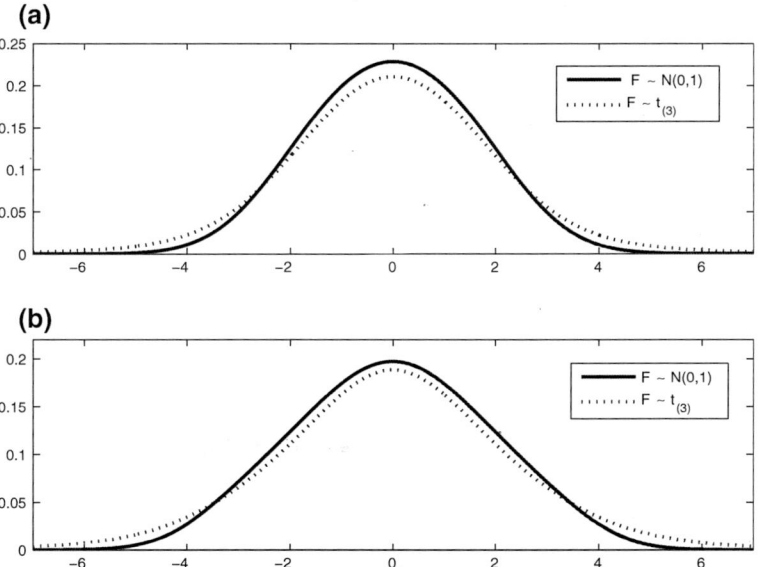

Fig. 4.2 Numerical approximation of the density of the C-convolution, g^C, when the marginal distribution H is standard Gaussian, the marginal distribution F is both standard Gaussian and Student's t with 3 degrees of freedom and the copula C is the Frank copula with level of dependence $\tau = 0.25$ (panel (**a**)) and $\tau = 0.75$ (panel (**b**))

The Gaussian Case

Let us assume that the copula C is gaussian, that is,

$$C(u, v) = \int_0^u \Phi\left(\frac{\Phi^{-1}(v) - \rho\Phi^{-1}(w)}{\sqrt{1 - \rho^2}}\right), \quad \rho \in (-1, 1),$$

where Φ is the standard normal cumulative distribution function.

Since $D_1 C(u, v) = \Phi\left(\frac{\Phi^{-1}(v) - \rho\Phi^{-1}(u)}{\sqrt{1-\rho^2}}\right)$, we get

$$H \overset{C}{*} F(t) = \int_0^1 \Phi\left(\frac{\Phi^{-1}(F(t - H^{-1}(w))) - \rho\Phi^{-1}(w)}{\sqrt{1 - \rho^2}}\right) dw$$

and

$$\hat{C}(u, v) = \int_0^u \Phi\left(\frac{\Phi^{-1}(F((H \overset{C}{*} F)^{-1}(v) - H^{-1}(w))) - \rho\Phi^{-1}(w)}{\sqrt{1 - \rho^2}}\right) dw.$$

Obviously, in general, $\hat{C}$ is no more gaussian, unless we assume that F and H are normally distributed. In fact, in such a case, $C(H(x), F(x))$ is the distribution function of a random vector (X, Y) normally distributed with margins H and F, respectively. It is a known fact that the random vector $(X, X + Y)$ is again normally distributed; hence, the distribution of $X + Y$ is normal and the associated copula $\hat{C}$ is of gaussian type. More precisely, if $H \sim N(\mu, \sigma^2)$ and $F \sim N(m, s^2)$, then $H \overset{C}{*} F \sim N(\mu + m, \sigma^2 + s^2 + 2\sigma s \rho)$ and

$$\hat{C}(u, v) = \int_0^u \Phi\left(\frac{\Phi^{-1}(v) - \hat{\rho}\Phi^{-1}(w)}{\sqrt{1 - \hat{\rho}^2}}\right),$$

where $\hat{\rho} = \frac{Cov(X, X+Y)}{\sigma_X \sigma_{X+Y}} = \frac{\sigma + s\rho}{\sqrt{\sigma^2 + s^2 + 2\sigma s\rho}}$.

The Elliptical Case

Let us assume now that the copula C is elliptical. We remind that C is the copula associated to a random vector (X, Y) having a bivariate elliptical distribution with margins G_1 and G_2, respectively. Formally,

$$C(u, v) = \int_{-\infty}^{G_1^{-1}(u)} \int_{-\infty}^{G_2^{-1}(v)} \sqrt{ac - b^2} g(as^2 + 2bst + ct^2) \, ds \, dt, \qquad (4.3)$$

where we assumed, without any loss of generality, that both G_1 and G_2 define distributions with zero mean. Recall that the parameters a, b, c are so that the symmetric matrix $\Sigma^{-1} = \begin{pmatrix} a & b \\ b & c \end{pmatrix}$ is positive definite. The elliptical copulas set contain gaussian copulas and Student's t copulas as particular cases. It is trivial to check that gaussian copulas can be recovered by considering

$$g(z) = \frac{1}{2\pi} e^{-\frac{z^2}{2}}.$$

Student's t with m degrees of freedom copulas are those copulas of type (4.3) with

$$g(z) = \frac{\Gamma(\frac{m+2}{2})}{\pi m \Gamma(\frac{m}{2})} \left(1 + \frac{z}{m}\right)^{-\frac{m+2}{2}}.$$

Since

$$D_1 C(u, v) = \sqrt{ac - b^2} \frac{1}{g_1(G_1^{-1}(u))} \int_{-\infty}^{G_2^{-1}(v)} g(a G_1^{-1}(u)^2 + 2b G_1^{-1}(u)t + ct^2)\, dt,$$

$$H \overset{C}{*} F(r) = \sqrt{ac - b^2} \int_0^1 \frac{1}{g_1(G_1^{-1}(w))}$$
$$\int_{-\infty}^{G_2^{-1}(F(r - H^{-1}(w)))} g(a G_1^{-1}(w)^2 + 2b G_1^{-1}(w)t + ct^2)\, dt\, dw$$

and

$$\hat{C}(u, v) = \sqrt{ac - b^2} \int_0^u \frac{1}{g_1(G_1^{-1}(w))}$$
$$\int_{-\infty}^{G_2^{-1}(F(H \overset{C}{*} F^{-1}(v) - H^{-1}(w)))} g(a G_1^{-1}(w)^2 + 2b G_1^{-1}(w)t + ct^2)\, dt\, dw.$$

If $H = G_1$ and $F = G_2$, then, $\hat{C}$ and $H \overset{C}{*} F$ are the copula function associated to $(X, X+Y)$ and the distribution of $X+Y$. It is a known fact, that $(X, X+Y)$ is again elliptically distributed. More precisely

$$G_1 \overset{C}{*} G_2(r) = \sqrt{ac - b^2} \int_0^1 \frac{1}{g_1(G_1^{-1}(w))}$$
$$\int_{-\infty}^{G_2^{-1}(G_2(r - H^{-1}(w)))} g(a G_1^{-1}(w)^2 + 2b G_1^{-1}(w)t + ct^2)\, dt\, dw =$$
$$= \sqrt{ac - b^2} \int_{+\infty}^{-\infty} \int_{-\infty}^{r-x} g(ax(w)^2 + 2bxt + ct^2)\, dt\, dx$$

and

$$\hat{C}(u, v) = \sqrt{ac - b^2} \int_0^u \frac{1}{g_1(G_1^{-1}(w))}$$

$$\int_{-\infty}^{(G_1 \overset{C}{*} G_2)^{-1}(v) - G_1^{-1}(w)} g(aG_1^{-1}(w)^2 + 2bG_1^{-1}(w)t + ct^2)\, dt dw =$$

$$= \sqrt{ac - b^2} \int_{-\infty}^{G_1^{-1}(u)} \int_{-\infty}^{(G_1 \overset{C}{*} G_2)^{-1}(v) - s} g(as^2 + 2bst + ct^2)\, dt dw =$$

$$= \sqrt{ac - b^2} \int_{-\infty}^{G_1^{-1}(u)} \int_{-\infty}^{(G_1 \overset{C}{*} G_2)^{-1}(v)} g(as^2 + 2bs(\hat{t} - s) + c(\hat{t} - s)^2)\, dt dw =$$

$$= \sqrt{ac - b^2} \int_{-\infty}^{G_1^{-1}(u)} \int_{-\infty}^{(G_1 \overset{C}{*} G_2)^{-1}(v)} g((a + c - 2b)s^2 + 2\hat{t}s(b - c) + c\hat{t}^2)\, dt dw$$

and this is of the same type as (4.3) with associated matrix $\begin{pmatrix} a + c - 2b & b - c \\ b - c & c \end{pmatrix}$.

4.2 Processes with Dependent Increments: Construction and Simulation

We are now ready to apply the C-convolution technique to build stochastic processes.

Let X_i with marginal distribution H_i, Y_{i+1} with distribution F_{i+1} and C_i be the copula associated to (X_i, Y_{i+1}). Then, we may recover the distribution of $X_{i+1} = X_i + Y_{i+1}$ as

$$H_{i+1}(x) = \int_0^1 D_1 C_i \left(w, F_{i+1}(x - H_i^{-1}(w)) \right) dw.$$

and the copula $C_{i,i+1}(u, v)$ associated to (X_i, X_{i+1}) as

$$C_{i,i+1}(u, v) = \int_0^u D_1 C_i \left(w, F_{i+1}(H_{i+1}^{-1}(v) - H_i^{-1}(w)) \right) dw.$$

Then, the copula $C_{i,i+1}(u, v)$ can be applied in the DNO approach to construct a Markov process. Notice that, unlike in the classical approach presented in Sect. 3.1.4, here we are free to specify the distribution of the starting level X_0 and the distributions of the increments of the process. Then the distributions of the levels are automatically determined.

The simulation of Markov processes with dependent increments is based on the technique of conditional sampling, as described in the quasi-algorithm reported below. The input is given by a sequence of distributions of increments that for the sake of simplicity we assume stationary $F_i = F$ and a temporal dependence structure that we consider stationary as well, $C_{X_i, Y_{i+1}}(u, v) = C(u, v)$. We also assume $X_0 = 0$. We describe a procedure to generate a iteration of a n-step trajectory.

1. $i = 1$
2. Generate u from the uniform distribution
3. Compute $X_i = F^{-1}(u)$
4. Use conditional sampling to generate v from $D_1 C(u, v)$
5. Compute $Y_{i+1} = F^{-1}(v)$
6. $X_{i+1} = X_i + Y_{i+1}$
7. Compute the distribution $H_{i+1}(t)$ by C-convolution
8. Compute $u = H_{i+1}(X_{i+1})$
9. $i = i+1$
10. If $i < n + 1$ go to step 4, Else End.

4.3 C-Convolution-Based Autoregressive Processes

A possible application of the C-convolution-based processes is in the study of the behavior of an autoregressive process of order 1 (AR(1) process) $X_t = \phi X_{t-1} + \varepsilon_t$ when X_{t-1} and ε_t are not independent as in the standard case but linked by some copula C_t. For a detailed discussion on autoregressive processes, we refer the reader to the manuals of Hamilton (1994) and of Brockwell and Davis (1991).

Recall that if the copula C_t is the independent copula, that is, $C_t(u, v) = uv$, the C-convolution coincides with the standard convolution and we obtain the standard AR(1) process. In this section we consider a C-convolution-based first-order autoregressive process, C-AR(1), by imposing a dependence structure between X_{t-1} and ε_t. The distribution of $X_t = \phi X_{t-1} + \varepsilon_t$ is given by the C-convolution between the distribution of ϕX_{t-1} and the distribution of ε_t. Suppose that X_1 has distribution F_1. Then, the distribution of X_t is

$$F_t(x_t) = F_{\phi X_{t-1}} \overset{C_t}{*} F_{\varepsilon_t}(x_t) = \int_0^1 D_1 C_t(w, F_{\varepsilon_t}(x_t - \phi F_{t-1}^{-1}(w))) dw, \quad t = 2, 3, \ldots$$

Moreover, the dependent structure between two subsequent observations is

$$C_{X_{t-1}, X_t}(u, v) = \int_0^u D_1 C_t(w, F_{\varepsilon_t}(F_t^{-1}(v) - \phi F_{t-1}^{-1}(w))) dw, \quad t = 2, 3, \ldots$$

4.3.1 The Gaussian Case

In most cases, the above integrals cannot be expressed in closed form and they have to be evaluated numerically. Simulations of a C-AR(1) process is not simple in general. However, as seen in Sect. 4.1.1, the Gaussian family is closed under C-convolution. In order to use this fact, suppose that the following conditions hold:

1. the initial distribution is Gaussian

$$X_1 \sim N(\mu_1, \sigma_1)$$

    and the distribution of innovations is Gaussian and stationary

$$\varepsilon_t \sim N(0, \sigma_\varepsilon);$$

2. the copula function linking X_{t-1} and ε_t is Gaussian and stationary, i.e., $C_t(u, v) = G(u, v; \rho)$, where $G(\cdot, \cdot; \rho)$ is the Gaussian copula with parameter ρ.

In this framework, by iterating the previous result to our C-AR(1) process we get

$$X_t \sim N(\mu_t, V_t^2), \tag{4.4}$$

where

$$\mu_t = \mathbb{E}[X_t] = \phi^{t-1}\mu_1, \quad t = 2, \ldots, \tag{4.5}$$

$$V_t^2 = Var(X_t) = \phi^{2(t-1)}V_1^2 + \left(\sum_{i=1}^{t-1}\phi^{2(i-1)}\right)\sigma_\varepsilon^2 + 2\rho\sigma_\varepsilon\sum_{i=1}^{t-1}\phi^{2i-1}V_{t-i}, \quad t = 2, \ldots \tag{4.6}$$

Moreover, the copula between X_{t-1} and X_t is Gaussian with parameters

$$\rho_{X_{t-1}, X_t} = \frac{\phi V_{t-1} + \rho\sigma_\varepsilon}{V_t}, \quad t = 2, \ldots,$$

where $V_1 = \sigma_1$.

Notice that the stationarity conditions on F_ε and C_t are not necessary to preserve the normality of the distribution of X_t.

An interesting question is to study the behavior of μ_t and V_t^2 when $t \to +\infty$. As regards μ_t we have as $t \to +\infty$

$$\mu_t \longrightarrow \begin{cases} 0, & \text{if } |\phi| < 1; \\ (+\infty)(\mu_1), & \text{if } \phi = 1. \end{cases}$$

The limiting behavior of the standard deviation V_t depends on ϕ. We distinguish the case where ϕ is strictly less than 1 in absolute value (stationary case in the standard AR(1)) from the case where ϕ is equal to 1 (unit root case in the standard AR(1)).

1. $|\phi| < 1$. $V_t = \left(\phi^2 V_{t-1}^2 + \sigma_\varepsilon^2 + 2\rho\phi\sigma_\varepsilon V_{t-1}\right)^{1/2} = h(V_{t-1})$ is a first-order nonlinear difference equation. We know that $V_t \to \tilde{V}$ as $t \to +\infty$ if $|h'(V)| < 1$ for V sufficiently close to $\tilde{V}$ and $\tilde{V} = h(\tilde{V})$. First, we compute $\tilde{V}$. By some algebra

$$\tilde{V} = \frac{\sigma_\varepsilon(\rho\phi + \left(\rho^2\phi^2 + 1 - \phi^2\right)^{1/2})}{1 - \phi^2}.$$

Moreover, it is not hard to prove that the condition $|h'(V)| < 1$ is satisfied for all $\rho \in (-1, 1)$.

2. $\phi = 1$. In this case with the same approach we obtain

$$V_t \longrightarrow \begin{cases} -\frac{\sigma_\varepsilon}{2\rho}, & \text{if } \rho \in (-1, 0); \\ +\infty, & \text{otherwise.} \end{cases}$$

The dependence structure between X_t and X_{t+k} can be determined by using the product copula operator. In fact, by $*$ product $G(u, v; \rho_1) * G(u, v; \rho_2) = G(u, v; \rho_1\rho_2)$. Therefore, since

$$C_{X_t, X_{t+k}} = C_{X_t, X_{t+1}} * C_{X_{t+1}, X_{t+2}} * \cdots * C_{X_{t+k-1}, X_{t+k}}$$

the copula between X_t and X_{t+k} is gaussian with parameter

$$\rho_{X_t, X_{t+k}} = \prod_{s=0}^{k-1} \frac{\phi V_{t+s} + \rho\sigma_\varepsilon}{V_{t+s+1}},$$

which represents the autocorrelation function of a C-AR(1) process.

We wonder if $\rho_{X_t, X_{t+k}}$ tends to zero as the lag k approaches to infinity. We study the case where $\phi = 1$. We know that if $|\rho_{X_t, X_{t+1}}| \le H < 1$ then $\rho_{X_t, X_{t+k}} \to 0$ as $k \to +\infty$. But $\rho_{X_t, X_{t+1}} = \frac{V_t + \rho\sigma_\varepsilon}{V_{t+1}}$ and $V_t \to -\frac{\sigma_\varepsilon}{2\rho}$ as $t \to +\infty$ means that there exists $\tilde{t}$ such that for any $t < \tilde{t}$ we have $V_t \le -\frac{\sigma_\varepsilon}{2\rho} = \frac{\sigma_\varepsilon}{2|\rho|}$. Therefore

$$\left| \frac{V_t + \rho\sigma_\varepsilon}{V_{t+1}} \right| \le \frac{V_t + |\rho|\sigma_\varepsilon}{V_{t+1}} \le \frac{\frac{\sigma_\varepsilon}{2|\rho|} + \rho\sigma_\varepsilon}{V_{t+1}}.$$

Moreover, since $V_{t+1} = V_1^2 + t\sigma_\varepsilon^2 + \sum_{i=1}^t V_{t+1-i} \ge V_1^2 + t\sigma_\varepsilon^2$ we can write

$$\left| \frac{V_t + \rho\sigma_\varepsilon}{V_{t+1}} \right| \le \frac{\frac{\sigma_\varepsilon}{2|\rho|} + \rho\sigma_\varepsilon}{V_1^2 + t\sigma_\varepsilon^2},$$

which approaches to zero when $t \to +\infty$. This implies that there exists $\tilde{\tilde{t}}$ such that if $t > \tilde{\tilde{t}}$

$$\frac{\frac{\sigma_\varepsilon}{2|\rho|} + \rho\sigma_\varepsilon}{V_1^2 + t\sigma_\varepsilon^2} \le H < 1.$$

This yields the result.

4.3.2 An Alternative Representation

The dependence structure of the C-convolution-based autoregressive process can be obtained modeling the innovations ε_t as $\varepsilon_t = \alpha_t X_{t-1} + u_t$, where $(u_t) \overset{i.i.d.}{\sim} N(0, \sigma_u)$ and X_{t-1} is independent of u_t for all $t \ge 2$. α_t is a time-dependent autoregressive coefficient, whose expression is consistent with the fact that the correlation coefficient between X_{t-1} and ε_t is equal to ρ for all t. We have

$$Cov(X_{t-1}, \varepsilon_t) = \mathbb{E}[X_{t-1}\varepsilon_t] = \mathbb{E}[\alpha_t X_{t-1}^2 + X_{t-1}u_t] = \alpha_t V_{t-1}^2,$$

then, α_t must satisfy the condition

$$\rho = \frac{\alpha_t V_{t-1}^2}{V_{t-1}\sigma_{\varepsilon_t}} = \frac{\alpha_t V_{t-1}}{\left(\alpha_t^2 V_{t-1}^2 + \sigma_u^2\right)^{1/2}},$$

since $\sigma_{\varepsilon_t}^2 = \alpha_t^2 V_{t-1}^2 + \sigma_u^2$. We obtain

$$\alpha_t = \frac{\rho\sigma_u}{V_{t-1}(1 - \rho^2)^{1/2}}.$$

The variance of the innovation is then

$$\sigma_{\varepsilon_t}^2 = \alpha_t^2 V_{t-1}^2 + \sigma_u^2 = \frac{\rho^2\sigma_u^2}{V_{t-1}^2(1 - \rho^2)} V_{t-1}^2 + \sigma_u^2 = \frac{\sigma_u^2}{1 - \rho^2},$$

which is constant for all t. It follows that the C-AR(1) process $X_t = \phi X_{t-1} + \varepsilon_t$ can be equivalently written

$$X_t = \phi_t X_{t-1} + u_t,$$

where $\phi_t = 1 + \alpha_t = 1 + \frac{\rho}{(1-\rho^2)^{1/2}} \frac{\sigma_u}{V_{t-1}}$. As for the limiting behavior of the autoregressive coefficient, we distinguish two cases according to the value of ϕ. If $|\phi| < 1$

$$\phi_t \longrightarrow 1 + \frac{\rho}{(1-\rho^2)^{1/2}} \frac{\sigma_u(1 - \phi^2)}{\sigma_\varepsilon(\rho\phi + (\rho^2\phi^2 + 1 - \phi^2)^{1/2})},$$

since $V_t \longrightarrow \frac{\sigma_\varepsilon(\rho\phi + (\rho^2\phi^2 + 1 - \phi^2)^{1/2})}{1 - \phi^2}$, whereas if $\phi = 1$ and $\rho < 0$ we have

$$\phi_t \to 1 - 2\rho^2, \quad t \to +\infty,$$

since $V_{t-1} \to -\frac{\sigma_\varepsilon}{2\rho} = -\frac{\sigma_u}{2\rho(1-\rho^2)^{1/2}}$.

On the other hand, we can consider the opposite case where the correlation coefficient is time-dependent but $\alpha_t = \alpha$ for all t. In this framework, the time-dependent correlation coefficient is given by

$$\rho_t = \frac{\alpha V_{t-1}}{\sqrt{\sigma_u^2 + \alpha^2 V_{t-1}^2}}$$

whereas the expression of the variance of X_t as a function of α is $V_t^2 = \phi^2 V_{t-1}^2 + \sigma_\varepsilon^2 + 2\phi\alpha V_{t-1}^2$.

4.4 Simulation Study

In this section, we present a Monte Carlo simulation study to compare some properties of a gaussian C-AR(1) process with a standard (and gaussian) AR(1) process. In particular we are interested in capturing the impact of the dependence between X_{t-1} and ε_t on the autocorrelation function and on the behavior of the Dickey–Fuller unit roots test.

4.4.1 Simulation Algorithm for a C-AR(1) Process

The simulation of a gaussian C-AR(1) process may be obtained by applying the technique of conditional sampling as described in the quasi-algorithm reported below. The input is given by a sequence of gaussian distributions of innovations that we assume stationary, $F_{\varepsilon_t} \sim N(0, \sigma_\varepsilon)$, and a temporal dependence structure that we consider gaussian and stationary as well, $C_{X_{t-1}, \varepsilon_t}(u, v) = G(u, v; \rho)$. We also assume $X_1 \sim N(\mu_1, V_1)$. The procedure to generate a trajectory of n points from the C-AR(1) process is the following:

1. $t = 1$.
2. Generate u from a uniform distribution.
3. Use conditional sampling to generate v from $D_1 C(u, v)$.
4. Compute $\varepsilon_{t+1} = F_\varepsilon^{-1}(v)$.
5. Compute $X_{t+1} = \phi X_t + \varepsilon_{t+1}$.
6. Compute the distribution of X_{t+1}, $F_{t+1}(y)$, by (4.4)–(4.6).
7. Compute $u = F_{t+1}(X_{t+1})$.
8. $t = t + 1$.
9. If $t < n + 1$ go to step 4, else End.

With this algorithm, we can study the autocorrelation function through a Monte Carlo simulation. We generate 5000 trajectories of a gaussian C-AR(1) process with two different values of the autoregressive coefficient $\phi = 0.8$ and $\phi = 0.95$ and three different values of the gaussian copula parameter $\rho = -0.5, 0.5, 0.8$. We compute

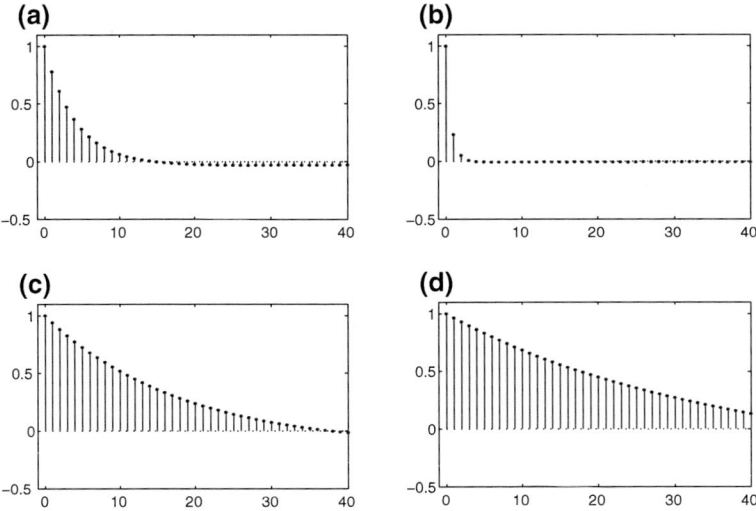

Fig. 4.3 Monte Carlo autocorrelation function. **a** Standard AR(1) with $\phi = 0.8$. Convolution-based AR(1) with $\phi = 0.8$ and **b** $\rho = -0.5$; **c** $\rho = 0.5$; **d** $\rho = 0.8$

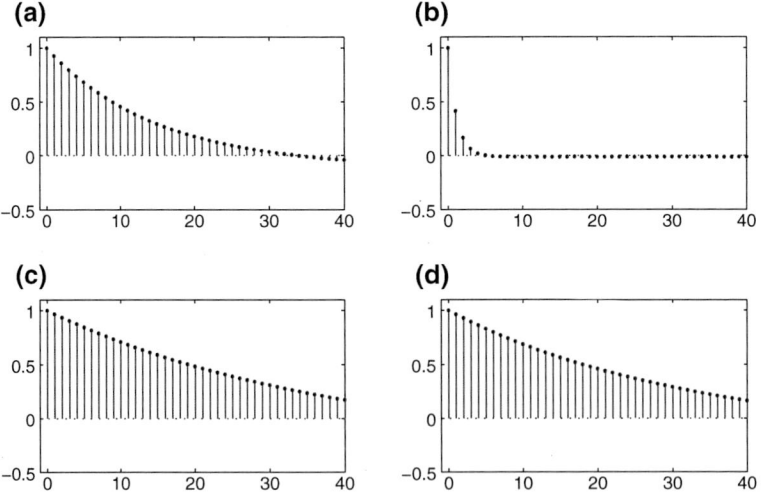

Fig. 4.4 Monte Carlo autocorrelation function. **a** Standard AR(1) with $\phi = 0.95$. Convolution-based AR(1) with $\phi = 0.95$ and **b** $\rho = -0.5$; **c** $\rho = 0.5$; **d** $\rho = 0.8$

the average autocorrelation function and we report the dynamics in Figs. 4.3 and 4.4. For the sake of comparison, we also display the autocorrelation function of the standard AR(1) process with $\phi = 0.8$ and $\phi = 0.95$. As we can observe the presence of negative correlation between X_{t-1} and ε_t pushes the process to the absence of memory as if it behaved as a moving average process of order 1 in the case of $\phi = 0.8$ and of order 2 in the case of $\phi = 0.95$. Incidentally, let us observe that a correlation of -0.5 is rather unrealistic in observed time series. When ρ is positive, the autocorrelation function decreases much more slowly than in the standard AR(1) model.

4.4.2 Small Sample Properties of OLS Estimator

In this section, we study the small samples properties of OLS estimator of the autoregressive coefficient ϕ for a C-AR(1) model for a number of different levels of dependency between X_{t-1} and ε_t. Currently asymptotic results are not available and we do not know the limit distribution of this estimator. Our Monte Carlo simulation is based on the algorithm introduced in the last subsection. We generate 5000 trajectories of 250 points of twelve different models obtained by choosing four values of the autoregressive coefficient ϕ, that are, 0.5, 0.8, 0.95, and 0.99 and six values of the correlation coefficient ρ, that are, -0.10, -0.05, -0.03, 0.03, 0.05, 0.10. For each trajectories we compute the OLS estimator $\hat{\phi}_n$ of ϕ given by

$$\hat{\phi}_n = \frac{\sum_{t=2}^{n} x_t x_{t-1}}{\sum_{t=2}^{n} x_{t-1}^2}.$$

Table 4.1 reports the results of the simulation. We can infer that for negative values of the correlation coefficient ρ the OLS estimator underestimates the parameter systematically. It is quite clear the presence of a negative bias which depends on ρ. The accuracy of the estimate is good and improves as ϕ grows and as ρ approaches zero. As expected, for positive values of ρ the estimator has a positive bias and in general the estimates are slightly more accurate. The histograms of the estimates for two simulated cases are reported in Figs. 4.5 and 4.6

4.4.3 Dickey–Fuller Unit Root Test

An interesting topic to investigate is how the Dickey–Fuller unit root test (Dickey and Fuller 1979; Hamilton 1994) performs in the case of a *unit root* C-AR(1) process. In other words, our data generating process is a modified unit root process $X_t = X_{t-1} + \varepsilon_t$ where the state variable X_{t-1} and the disturbance ε_t are linked by a gaussian copula with correlation coefficient ρ. We will show that the test does not perform well at all, when rho is just slightly negative.

Table 4.1 Least squares estimates by Monte Carlo simulation and relative root mean square error (in percentage value). The second number in brackets indicates the percentage of underestimates

	$\rho = -0.10$	$\rho = -0.05$	$\rho = -0.03$
$\phi = 0.5$	$\hat{\phi}_n = 0.4082$ (11.11%)(95.16%)	$\hat{\phi}_n = 0.4502$ (7.51%)(79.94%)	$\hat{\phi}_n = 0.4702$ (6.39%)(69.40%)
$\phi = 0.8$	$\hat{\phi}_n = 0.7258$ (8.62%)(97.06%)	$\hat{\phi}_n = 0.7628$ (5.54%)(81.52%)	$\hat{\phi}_n = 0.7754$ (4.76%)(71.60%)
$\phi = 0.95$	$\hat{\phi}_n = 0.9003$ (5.74%)(98.78%)	$\hat{\phi}_n = 0.9247$ (3.57%)(85.02%)	$\hat{\phi}_n = 0.9324$ (2.99%)(75.03%)
$\phi = 0.99$	$\hat{\phi}_n = 0.9537$ (4.20%)(99.84%)	$\hat{\phi}_n = 0.9746$ (2.55%)(90.62%)	$\hat{\phi}_n = 0.9769$ (2.10%)(80.84%)
	$\rho = 0.03$	$\rho = 0.05$	$\rho = 0.10$
$\phi = 0.5$	$\hat{\phi} = 0.5208$ (5.76%)(32.60%)	$\hat{\phi} = 0.5377$ (6.57%)(23.18%)	$\hat{\phi} = 0.5788$ (9.39%)(6.70%)
$\phi = 0.8$	$\hat{\phi} = 0.8104$ (3.93%)(35.54%)	$\hat{\phi} = 0.8214$ (4.26%)(26.12%)	$\hat{\phi} = 0.8459$ (5.69%)(9.82%)
$\phi = 0.95$	$\hat{\phi} = 0.9513$ (2.15%)(39.54%)	$\hat{\phi} = 0.9563$ (2.11%)(31.22%)	$\hat{\phi} = 0.9664$ (2.46%)(15.92%)
$\phi = 0.99$	$\hat{\phi} = 0.9865$ (1.42%)(48.94%)	$\hat{\phi} = 0.9888$ (1.33%)(39.72%)	$\hat{\phi} = 0.9923$ (1.26%)(27.04%)

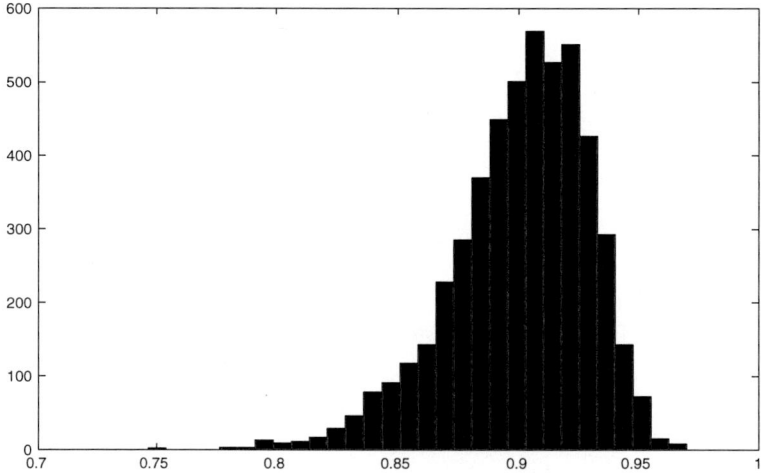

Fig. 4.5 Histogram of Monte Carlo least square estimates of the autocorrelation coefficient when the true model is C-AR(1) with $\phi = 0.95$ and $\rho = -0.10$

There are many versions of the Dickey–Fuller test and each of them has its own critical value which depends on the size of the sample. In each case, the null hypothesis is that there is a unit root, i.e., the true process is a unit root process. The tests have low statistical power in that they often cannot distinguish between true unit root processes and near unit root processes.

The intuition behind the test is as follows. If the time series $(X_t)_t$ is stationary, then it has a tendency to return to a constant mean. Therefore, large values will tend to be followed by smaller values and small values by larger values. Accordingly, the level of the series will be a significant predictor of next period's change, and will have a negative coefficient. Suppose the true process is a unit root C-AR(1) process with a correlation coefficient between X_{t-1} and ε_t given by ρ, and suppose that we estimate by ordinary least squares (OLS) the autoregressive coefficient based on a sample of size n, say $\hat{\phi}_n$. We apply two different versions of the Dickey–Fuller test. We denote by $df_{1,n}$ and $df_{2,n}$ the test statistics of the test whose expressions are given by

$$df_{1,n} = n(\hat{\phi}_n - 1),$$

and

$$df_{2,n} = \frac{\hat{\phi}_n - 1}{\hat{\sigma}_{\hat{\phi}_n}},$$

where $\hat{\sigma}_{\hat{\phi}_n}$ is the standard error of the OLS estimator. There exist exact asymptotic distributions of $df_{1,n}$ and $df_{2,n}$, whereas in small samples only approximate distributions for different values of the sample size are available (Fuller 1976).

These tables report the critical values of the distribution of $df_{1,n}$ and $df_{2,n}$ for which, if the test statistics is less or equal to them, at a fixed level of significance, the null hypothesis of unit root can be rejected. For example, for $df_{1,n}$ the critical value is -13.6 at 1% level of significance and -8 and at 5% level of significance if

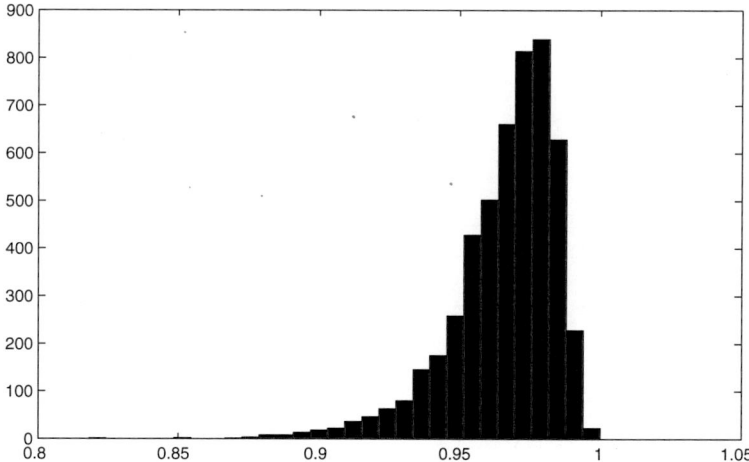

Fig. 4.6 Histogram of Monte Carlo least square estimates of the autocorrelation coefficient when the true model is C-AR(1) with $\phi = 0.95$ and $\rho = 0.10$

Table 4.2 Percentage of rejection of the null hypothesis in which the true process is a unit root process using the two Dickey–Fuller test statistics

		$\rho = -0.03$	$\rho = -0.05$	$\rho = -0.30$
$\alpha = 1\%$	$df_{1,n}$	1.50%	2.14%	9.42%
	$df_{2,n}$	1.56%	2.16%	9.52%
$\alpha = 5\%$	$df_{1,n}$	7.68%	11.10%	33.92%
	$df_{2,n}$	7.78%	11.24%	33.68%

the sample size is $n = 250$. As regards $df_{2,n}$ the critical values are -2.58 and -1.95 respectively (Fig. 4.6).

Our purpose is to capture the impact of the presence of ρ on the performance of the Dickey–Fuller test. In this regard, we simulate 5000 trajectories of 250 points of a gaussian unit root C-AR(1) process with three different levels of correlation $\rho = -0.03, -0.05, -0.10$ by using the algorithm presented in Sect. 4.1.1. For each simulated trajectory we compute $df_{1,n}$ and $df_{2,n}$. We compare the simulated values of the test statistics with the critical values at 1 and 5% level of significance of the standard Dickey–Fuller test. Table 4.2 reports the results. As we can argue, in the presence of a negative correlation between the state variable and the disturbance, the test statistics is characterized by strong negative skewness greater than in the standard case. In other words, we need a value of the test statistics much more negative than in the independent case since the presence of negative correlation "retrieves a bit of stationary." This effect is stronger with increasing level of significance.

References

Brockwell, P. J., & Davis, R. A. (1991). *Time series. Theory and methods*. Springer series in statistics. New York: Springer.

Cherubini, U., Mulinacci, S., & Romagnoli, S. (2011). A copula-based model of speculative price dynamics in discrete time. *forthcoming in Journal of Multivariate Analysis*.

Cherubini, U., Gobbi, F., Mulinacci, S., & Romagnoli, S. (2012). *Dynamic copula methods in finance*. New York: Wiley.

Dickey, D. A., & Fuller, W. A. (1979). Distribution of the estimators for autoregressive time series with a unit root. *Journal of the American Statistical Association, 74*, 427–431.

Fuller, W. A. (1976). *Introduction to statistical time series*. New York: Wiley.

Hamilton, J. D. (1994). *Time series analysis*. Princeton: Princeton University Press.

Chapter 5
Application to Interest Rates

5.1 Nonlinear Behavior in Interest Rates: A Review

There is a large literature investigating the nonlinear dynamics of the short-term rate. It mainly dates back to the last decade of the last century. Most of this literature was about persistence or mean reversion, linearity or nonlinearity, Gaussian or non-Gaussian innovations. Moreover, it is all about extensions and distortions of the linear AR(1) model, that is the subject addressed in this book. It is then the appropriate application to show how our approach works in practice, and maybe to stimulate new research on the subject. This would also be particularly welcome in view of the puzzles offered by the interest rate markets in the aftermath of the crisis of 2007–2008, when the interest rates began to decrease dramatically, eventually reaching the negative territory.

The research on the dynamics of the interest rates began between the 1980s and the 1990s, with the first task of estimating and validating the continuous time models that had been proposed for the determination of the term structure and the pricing of bonds and interest rate derivatives. The models, originally defined in continuous time, can be transposed in the discrete time in the general form

$$r_t - r_{t-1} = a + br_{t-1} + \epsilon_t \tag{5.1}$$

where r_t denotes the time series of the short-term interest rate, a and b are parameters and ϵ_t is a sequence of random innovations. The models proposed in the term structure literature based on the spot rate typically include mean reversion, corresponding to a parameter $b < 1$, and different assumptions about the distribution of the innovations ϵ_t. In the two most famous models, the innovations are Gaussian or heteroschedastic. In Vasicek (1977) the innovations are Gaussian and the conditional distribution of the future spot rates is normal. In Cox et al. (1985) (CIR) the innovation is not Gaussian, and the variance of the rate grows linearly with the level of interest rates: in this case, the conditional distribution of future spot rates is non-central chi-square. The first

© The Author(s) 2016
U. Cherubini et al., *Convolution Copula Econometrics*,
SpringerBriefs in Statistics, DOI 10.1007/978-3-319-48015-2_5

econometric works investigating whether these theoretical models where borne out by the data, and which of them provided the best fit were carried out by Marsh and Rosenfeld (1983) and Chan et al. (1992).

Beyond the task of selecting models, further research was directed to analyze the presence of nonlinearities in the dynamics of interest rates, both in the drift and the volatility of the process. Several indications of nonlinear dynamics were found, and different approaches were suggested to model these nonlinearities. Among all these papers we remind of the switching regime model proposed by Hamilton (1988), the analysis of asymmetric responses to shocks of different sign found in Kozicki (1994) and the autoregressive threshold model applied in Pfann et al. (1996).

Concerning the main issues investigated in the literature, a crucial question is whether the shocks to interest rates are permanent or transitory, that is the topic of mean reversion. In most of the literature, shocks to the short-term interest rates are found to be highly persistent, apart from nonlinear effects that will be discussed below. The non-stationary behavior is also consistent with the usual evidence that most of the changes in the term structure consist of parallel shifts. In fact, according to the "expectations hypothesis", long-term interest rates are equal to the average of future expected short rates. This implies that if the shock reaching the short-term rates is permanent, the shock to the average, that is the long-term rate, should be the same. If instead the shock were mean reverting, the impact on long-term rates should be smoothed out, and the impact on the term structure should be a change of slope, that is a "twist". The usual evidence is that "twists" account for a much smaller percentage of changes in the term structure.

Concerning nonlinearities, the main finding is that the mean reversion feature may change with the level of the interest rates, so that shocks are found to be highly persistent when the interest rates are low, while there is evidence of mean reversion only when interest rates are very high. The behavior is then in a sense asymmetric because the interest rates quickly revert to the mean when they are in double digits, while when they are very low they crawl around zero, or below zero as in these days. There is also substantial evidence of nonlinearity in the volatility figure, but this is mainly used to model non-Gaussian innovations.

Another issue that is usually found in all the empirical work on interest rate dynamics, and that in some sense is linked to the nonlinearity effects reported above is the parameter instability and the presence of structural breaks. This is quite typical because short-term interest rates are strongly dependent on the monetary policy. In fact, in the typical literature, that is mostly devoted to the US market, the break is found in the change in monetary policy that took place between 1979 and 1982, when the Federal Reserve switched its policy from the management of interest rates to the control of monetary aggregates. We are not aware, to the best of our knowledge, of any such analysis concerning the period of the recent crisis. In that period, monetary policy both in the US and Europe had to resort to nonstandard tools, such as the so-called "Quantitative Easing" programs. It would then be a very interesting and original task to start an analysis of the structural breaks and nonlinearities present in this period.

5.2 Nonlinear Drift Models

In Chap. 1 we proposed three models of nonlinear drift, that could be usefully applied to provide a contribution the literature above. We report here the models for the ease of the reader

- model 1: $Y_t = \alpha + \exp\left(-\beta Y_{t-1}\right) Y_{t-1} + \epsilon_t$
- model 2: $Y_t = \alpha + \exp\left(-\beta Y_{t-1}^2\right) Y_{t-1} + \epsilon_t$
- model 3: $Y_t = \alpha + \exp\left(-\beta(Y_{t-1} - \hat{Y})^2\right) Y_{t-1} + \epsilon_t$, with $\hat{Y} = 1\%$.

The models above represent a departure from the random walk dynamics that was found typical of interest rates movements in applied analysis, as well as from the linear mean reversion dynamics which is often used in theoretical models of the short-term rates. Figures 5.1, 5.2 and 5.3 report a Monte Carlo simulation which shows the changes in paths that could be generated by dynamic models like those proposed in this chapter. In the simulation we assume that $(\epsilon_t)_t \overset{i.i.d.}{\sim} N(0, \sigma)$. Moreover, we have set $\alpha = 0$ and $\sigma = 1.5\%$ and we have highlighted the effect of β on the dynamics. We see that all the models actually induce mean reversion and the path is described by a concave trajectory. Moreover, the path seems to converge towards something similar random walk when interest rates are sufficiently low. The trajectories are very similar, even though for model 1 the decrease looks more persistent, while for the other two models after a period of sudden decrease the path seems to float around a flat (or very slowly decreasing) level.

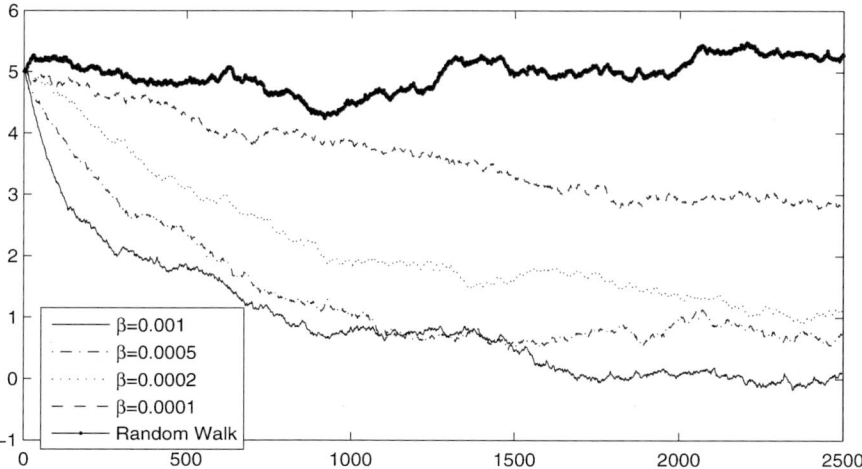

Fig. 5.1 Simulation of paths relative to model 1 for different level of the parameter β

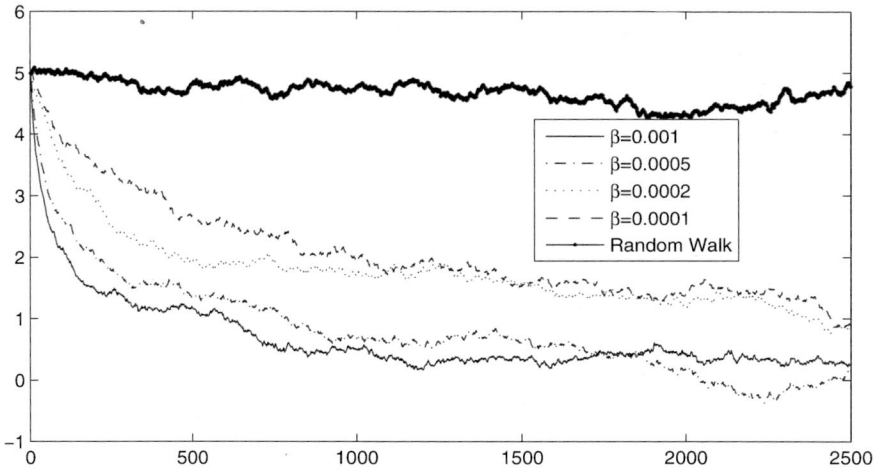

Fig. 5.2 Simulation of paths relative to model 2 for different level of the parameter β

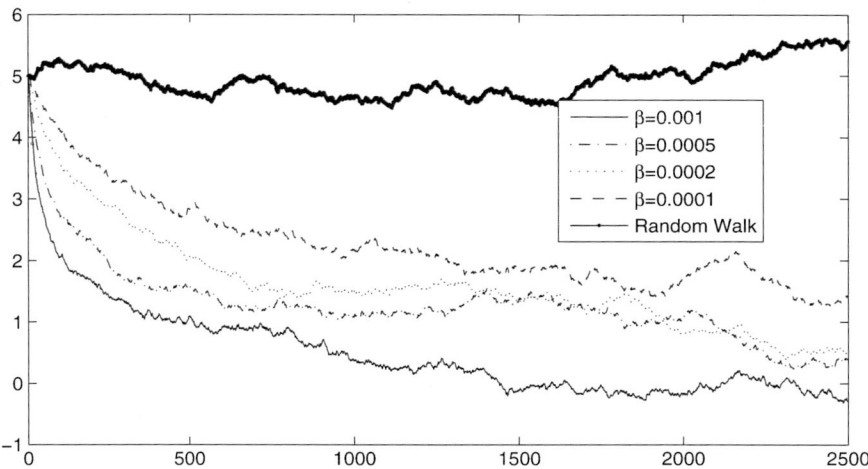

Fig. 5.3 Simulation of paths relative to model 3 for different level of the parameter β

5.3 An Analysis of the Short-Term Rate

It is now time to disclose the time series of short-term rates that we used as a reference from real world in Chap. 1, and to apply our convolution analysis to estimate the models above. Figure 5.4 reports the dynamics of this crucial economic variable, on a daily time series covering from January 2001 to May 2010, for a total of 2433 observations. Right from the visual inspection of the data,[1] it emerges clearly that the

---

[1] *Source* Thomson Reuters Datastream.

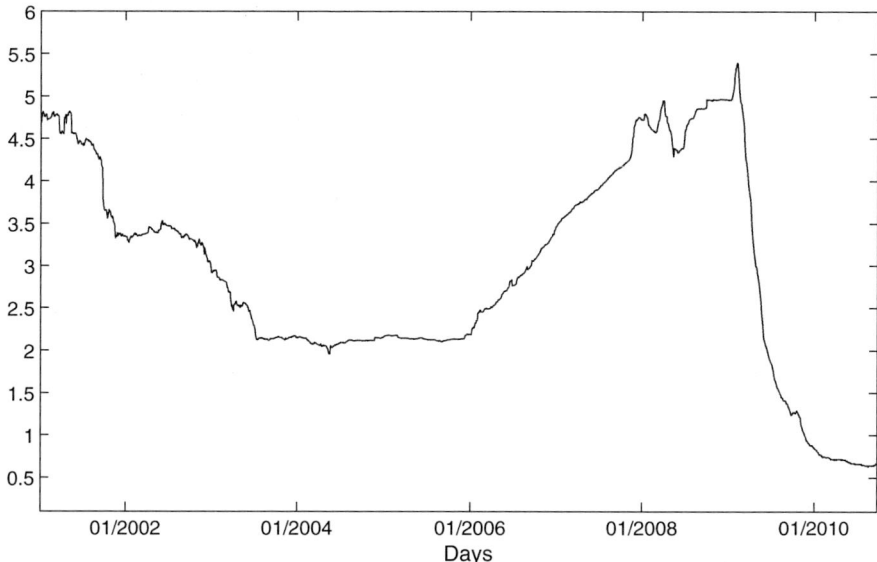

Fig. 5.4 Three month Euribor rate

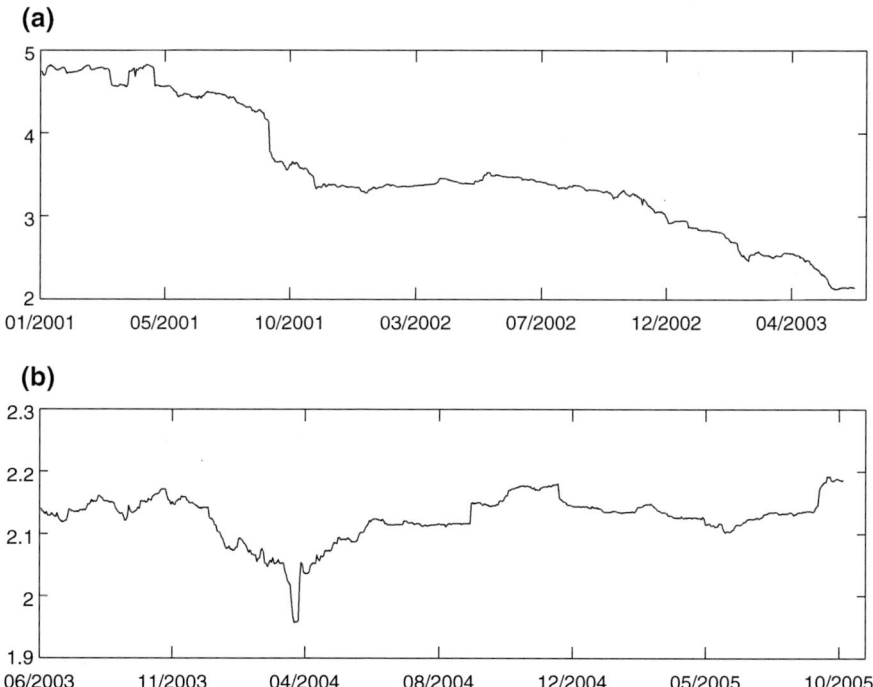

Fig. 5.5 Three month Euribor rate. **a** First period: from January 2001 to May 2003. **b** Second period: from June 2003 to October 2005

behavior of the short-term rate was quite different across the sample. In particular, we may identify four different periods (Figs. 5.5 and 5.6). The first period saw a steady decrease of the rate from 2001 to 2003, following the introduction of the Euro as a new currency and what was called the "convergence game", that is a general downward trend of the interest rates in all the European countries around the level of the most virtuous ones, particularly Germany. This was followed by a period of stationary floating around a level roughly above 2%, with marked variation and a deep fall in 2004. Then, following the worldwide business cycle, in the third period the drift turned positive, causing the interest rate to rise from 2 to 5% from end of 2005 to October 2008, at the end of "subprime" crisis in the US. In the last period, when the crisis migrated to Europe and joined with sovereign crisis, the interest rate dramatically changed its direction, reaching below the 1% barrier in less than two years.

We may now disclose that these four subperiods were actually those used in Chap. 1 to illustrate how to gauge the presence of a unit root using non-parametric methods, and namely the Kendall function of the association of the increments of the interest rate in each period and its level at the beginning of the period. We remind that we found evidence of mean reversion only in the third and fourth period, that is

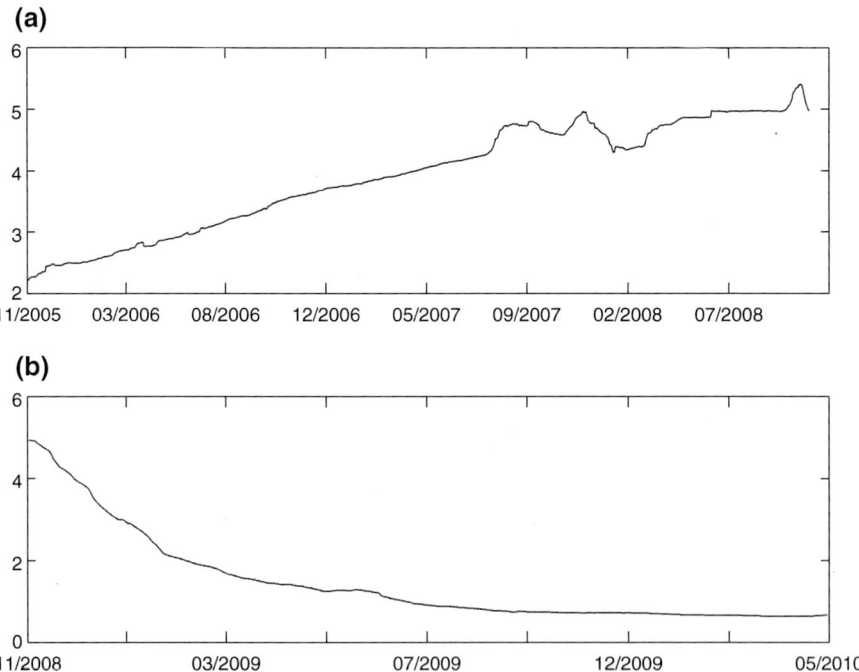

Fig. 5.6 Three month Euribor rate. **a** Third period: from November 2005 to October 2008. **b** Fourth period: from November 2008 to May 2010

from 2006 to the end of the sample, with the evidence of the mean reversion stronger in the last period, that is after 2008.

Given these preliminary results, we are now going to check if our model above, in its different specifications, can provide additional information on the mean reversion dynamics. Given those evident changes of dynamics, it makes sense to ask whether some nonlinearities are involved, or whether all these changes can be traced back to a random walk. Below, we are going to investigate if our models can be used to explain this dynamics. The questions that we are going to address are two. First, we ask whether there exists some parameter set that enables to explain the different dynamics in the subperiods. Second, if this is not true, we address the question if some nonlinearities are present at least in some of the subperiods.

5.4   Estimation and Results

In this section we present a maximum likelihood estimation of the parameters of the three models proposed above. Before doing that, however, those models must be extended in such a way that the standard linear mean reversion or random walk models could be nested in our estimation. It is quite clear that the random walk dynamics is already nested in our models in the case with $\psi(Y_{t-1}) = 1$. On the contrary, the linear mean reversion model is not nested, unless include one parameter more. We then use the general specification

$$Y_t = \alpha + \gamma \psi (Y_{t-1}) Y_{t-1} + \epsilon_t$$

where the scale parameter γ enables to take into account of a standard linear mean reverting AR(1) model when $\psi(Y_{t-1}) = 1$ and $\gamma < 1$.

We then perform the MLE estimation of the parameters α, β, γ and σ for the three models proposed. The method (Hamilton 1994) is based on the conditional distribution of Y_t given $Y_{t-1} = y_{t-1}$. In fact, if $(\epsilon_t)_t \overset{iid}{\sim} N(0, \sigma)$ then the conditional distribution of Y_t given $Y_{t-1} = y_{t-1}$ is given by

$$Y_t | y_{t-1} \sim N (\alpha + \gamma \psi (y_{t-1}) y_{t-1}, \sigma).$$

It is then easy to write the log-likelihood function of the time series generated by the three models proposed with different specifications for $\psi(y_{t-1})$.

Our estimation strategy was to perform first the linear model, setting $\beta = 0$. The results obtained were consistent with the nonparametric results quoted in Sect. 5.1. The model data showed mean reversion both in the third and the fourth subperiods, while confirming the evidence of a random walk dynamics in the first two.

Then, we applied our three models to explore nonlinearities in the data. The results are reported in Table 5.1. In all the models, we are particularly interested in checking if γ is statistically lower than 1 and if β is significantly different from 0.

Table 5.1 Maximum likelihood estimation of parameters. The asterisk denotes the parameters which are significantly different from zero at the 5% level. The double asterisk denotes when the parameter γ is significantly lesser than 1

Model 1

Parameter	Full data	Period 1	Period 2	Period 3	Period 4
$\hat{\alpha}$	-0.0011	-0.0155	$-$	-0.0065	0.0145^*
$\hat{\beta}$	-6.7333×10^{-5}	8.8365×10^{-4}	$-$	0.0012	-0.0021^*
$\hat{\gamma}$	0.9999	1.0065	$-$	1.0076	0.9780^{**}
$\hat{\sigma}$	0.0173^*	0.0240^*	$-$	0.0149^*	0.0110^*

Model 2

Parameter	Full data	Period 1	Period 2	Period 3	Period 4
$\hat{\alpha}$	-0.0014	-0.0114	$-$	-0.0021	0.0131^*
$\hat{\beta}$	-2.9146×10^{-6}	7.5925×10^{-4}	$-$	$1.1683 \times 10^{-4*}$	$-3.1808 \times 10^{-4*}$
$\hat{\gamma}$	1.0002	1.0032	$-$	1.0035	0.9813^{**}
$\hat{\sigma}$	0.0173^*	0.0240^*	$-$	0.0149^*	0.0110^*

Model 3

Parameter	Full data	Period 1	Period 2	Period 3	Period 4
$\hat{\alpha}$	-0.0016^*	-0.0114	-0.0030	0.0107^*	0.0118^*
$\hat{\beta}$	1.6509×10^{-5}	1.7937×10^{-4}	-0.0889^*	$1.1683 \times 10^{-4*}$	$-5.4839 \times 10^{-4*}$
$\hat{\gamma}$	1.0003	1.0004	1.0014	0.9985	0.9829^{**}
$\hat{\sigma}$	0.0173^*	0.0240^*	0.0049^*	0.0149^*	0.0110^*

The latter term controls for nonlinearity, while the former, if the hypothesis $\beta = 0$ is not rejected, provides a test of linear mean reversion versus a random walk model. A look at the first column of the table, where we report the estimation for the whole period, shows that the three models reject both the hypothesis of mean reversion and that of nonlinearity. In all the models, the estimates are consistent with a random walk model. The only difference is that in model 3 the constant is significantly negative, implying a negative drift.

But, as we said, the estimation of the whole sample did not look promising already from a look at the graph. So, the analysis of subperiods unveils more interesting results. In the first period the evidence of the three models completely agree with a unit root model. The drift is slightly negative, but not significant. The drift was significantly negative in the linear estimate. For the second period, the estimation failed to converge in model 1 and 2, while it provided evidence consistent with a random walk for model 3. Actually the nonlinear parameter turns out significantly negative, implying an explosive behavior, with a tendency to drift away faster for levels higher and lower than 1%. In preliminary estimates in which the γ parameter was set equal to 1, the nonlinearity parameter was found not statistically significant. In that case, the pure random walk model was also confirmed for models 1 and 2, for which in that case the estimation reached convergence. The pure random walk hypothesis then emerges as the largely dominant evidence, consistent with the nonparametric evidence of Chap. 1 and the linear model estimates.

While our models only provided a confirmation of standard linear model, they reveal important differences in the third and fourth period. The third period is the most interesting. We remind that for this period the linear model would estimate a mean reversion model, with mean reversion parameter equal to -0.0017 and significantly different from zero, and with a parameter of the lagged interest rate equal to 0.9983. In all our nonlinear models, instead, we find that γ is not significantly different from 1, while there is slight evidence of nonlinearity, even thought the parameter is very low, $\beta = 0.000163$. This produces values very close to the 1 for all the plausible values of the interest rate, but always strictly lower than 1. The model seems then consistent with some kind of long memory behavior, which is quite different from that predicted by the linear model. The only point of weakness of this evidence is that the parameters are not significant for model 1.

Finally, the fourth period offers another instance of our models. All the three models find that the parameter γ is significantly lower than 1, so that the evidence is strongly in favor of mean reversion. The parameter, ranging between 0.978 and 0.983, only slightly lower that the value estimated in the linear model, that is 0.987. The evidence in favor of mean reversion is highly significant, at the 1% probability level in all the models. There is also evidence of some nonlinearity: the β parameter is significantly different from zero, and it is negative. So, there is evidence that the parameter of the lagged variable is actually increasing with the level of the interest rate. However, the size of the parameter is so small that the mean reversion parameter can be considered constant across the relevant spectrum of values of the variable for all practical purposes: for interest rate values ranging from 0.50 to 5.50% the parameter remains the same up to the sixth digit.

We may then conclude that our convolution approach allows us to find evidence of nonlinear behavior in the drift of the short-term interest rates. In some instance, this nonlinearity may not have practical relevance, and the linear model can be maintained as a good representation of the dynamics of the variable. But there are cases in which the nonlinearity designs an alternative model with respect to the linear estimation. It is the case when the parameter of the lagged variable is kept very close to 1, but strictly lower than that, and with the mean reversion effect increasing, very smoothly, with higher levels of the short term rate. This would naturally give rise to a long memory effect in the dynamics of the variable. Apart from the convolution technique that was applied, then, our approach looks like a promising model to investigate the dynamics of the short rate. More research should be carried out on this topic, particularly with the new situation of negative interest rates: for example, models 1 and 2 would clearly behave in opposite directions with negative rates. For model 3 one could conceive to select a level below zero where the autoregressive parameter is set equal to 1, or else the level could be estimated from the data. It is only the beginning of the possible models that the convolution approach can implement, and we hope that the techniques explained in this book could be useful for this purpose.

References

Chan, K. C., Karolyi, G. A., Longstaff, F. A., & Sanders, A. B. (1992). An empirical comparison of alternative models of the short-term interest rate. *Journal of Finance, XLVII*(3).

Cox, J. C., Ingersoll, J., & Ross, S. (1985). A theory of the term structure of interest rates. *Econometrica, 53*, 385–407.

Hamilton, J. D. (1988). Rational expectations econometric analysis of changes in regime: An investigation of the term structure of interest rates. *Journal of Economic Dynamics and Control, 12*, 385–423.

Hamilton, J. D. (1994). *Time series analysis*. Princeton: Princeton University Press.

Kozicki, S. (1994). A nonlinear model of the term structure, Working Paper (Federal Reserve Board, Washington, DC).

Marsh, T. A., & Rosenfeld, E. R. (1983). Stochastic processes for interest rates and equilibrium bond prices. *Journal of Finance, 38*, 635–646.

Pfann, G. A., Schotman, P. C., & Tschemig, R. (1996). Nonlinear interest rate dynamics and implications for the term structure. *Journal of Econometrics, 74*, 149–176.

Vasicek, O. A. (1977). An equilibrium characterization of the term structure. *Journal of Financial Economics, 5*, 177–188.

Bibliography

Aas, K., Czado, C., Frigessi, A., & Bakken, H. (2009). Pair copula constructions of multiple dependence. *Insurance Mathematics and Economics, 44*(2), 182–198.

Amsler, C., Prokhorov, A., & Schmidt, P. (2010). Using copulas to model time dependence in stochastic frontier models, Working Paper.

Andrews, D. W. K. (1988). Law of large numbers for dependent non-identically distributed random variables. *Econometric Theory, 4*, 458–467.

Baglioni, A., & Cherubini, U. (2010). Marking-to-market government guarantees to financial systems: An empirical analysis of Europe, Working Paper.

Barbe, P., Genest, C., Ghoudi, K., & Rémillard, B. (1996). On Kendall's process. *Journal of Multivariate Analysis, 58*, 197–229.

Billingsley, P. (1986). *Probability and measure* (2nd ed.). New York: Wiley.

Block, H. W., & Ting, M. L. (1981). Some concepts of multivariate dependence. *Communications in Statistics - Theory and Methods, 10*, 749–762.

Block, H., Savits, T., & Shaked, M. (1982). Some concepts of negative dependence. *Annals of Probability, 10*, 765–772.

Blomqvist, N. (1950). On a measure of dependence between two random variables. *Annals of Mathematical Statistics, 21*, 593–600.

Bollerslev, T. (1986). Generalized autoregressive conditional heteroschedasticity. *Journal of Econometrics, 31*, 307–327.

Bollerslev, T. (1987). A conditional heteroschedasticity time series model for speculative prices and rates of return. *Review of Economics and Statistics, 69*, 542–547.

Cazzulani, L., Meneguzzo, D., & Vecchiato, W. (2001). Copulas: A statistical perspective with empirical applications, IntesaBCI Risk Management Department, Working Paper.

Cherubini, U., & Gobbi, F. (2013). A convolution-based autoregressive process. In F. Durante, W. Haerdle, & P. Jaworski (Eds.), *Workshop on copula in mathematics and quantitative finance*. Lecture notes in statistics-proceedings. Berlin/Heidelberg: Springer.

Cherubini, U., Luciano, E., & Vecchiato, W. (2004). *Copula methods in finance*. John Wiley series in finance. London: Wiley.

Cherubini, U., Gobbi, F., & Mulinacci, S. (2010). Semi parametric estimation and simulation of actively managed portfolios, Working Paper.

Choros, B., Ibragimov, R., & Permiakova, E. (2010). Copula estimation. In P. Jaworski, F. Durante, W. K. Haerdle, & T. Rychlik (Eds.), *Copula theory and its applications* (Vol. 198, pp. 77–91). Lecture notes in statistics-proceedings. Berlin/Heidelberg: Springer.

Clayton, D. G. (1978). A model for association in bivariate life tables and its application in epidemiological studies of familial tendency in chronic disease incidence. *Biometrika, 65*, 141–151.

Cont, R. (2001). Empirical properties of asset returns: Stylized facts and statistical issue. *Quantitative Finance, 1*, 223–236.

Davidson, R., & MacKinnon, J. (1993). *Estimation and inference in econometrics*. Oxford: Oxford University Press.

Davydov, Y. (1973). Mixing conditions for Markov chains. *Theory of Probability and Its Applications, 18*, 312–328.

Deheuvels, P. (1978). Caractérisation complète des Lois Extrèmes Multivariées et de la Convergence des Types Extrèmes. *Publications de l'Institut de Statistique de l'Université de Paris, 23*, 1–36.

Deheuvels, P. (1979). La function de dépendance empirique et ses propriétés. Un test non paramétriquen d'indépendace, *Acad. Roy. Belg. Bull. C1 Sci., 65*(5), 274–292.

Deheuvels, P. (1981). *A non parametric test for independence*. Université de Paris, Institut de Statistique.

Dickey, D. A. (1976). Estimation and hypothesis testing in non-stationary time series, Unpublished doctoral dissertation (Iowa State University, Ames, IA).

Dickey, D. A., & Fuller, W. A. (1981). Likelihood ratio statistics for autoregressive time series with a unit root. *Econometrica, 49*, 1057–1072.

Diebold, F. X., Gunther, T., & Tay, A. S. (1998). Evaluating density forecasts with applications to financial risk management. *International Economic Review, 39*, 863–883.

Diebold, F. X., Hahn, J., & Tay, A. S. (1999). Multivariate density forecast evaluation and calibration in financial risk management. *Review of Economics and Statistics, 81*(4), 661–673.

Domowitz, I., & White, H. (1982). Misspecified models with dependent observations. *Journal of Econometrics, 20*, 35–58.

Drouet-Mari, D., & Kotz, S. (2001). *Correlation and dependence*. London: Imperial College Press.

Durante, F., & Sempi, C. (2015). *Principles of copula theory*. Boca Raton: Chapman and Hall/CRC.

Durrleman, V., Nikeghbali, A., & Roncalli, T. (2000). Which copula is the right one?, Groupe de Recherche Opèrationelle, Cr édit Lyonnais, Working Paper.

Engle, R. F. (Ed.). (1996). *ARCH selected readings*. Oxford: Oxford University Press.

Engle, R. F. (2002). Dynamic conditional correlation. *Journal of Business & Economic Statistics, 20*(3), 339–350.

Engle, R. F., & Manganelli, S. (1999). CAViaR: Conditional autoregressive value at risk by regression quantiles, UCSD Department of Economics, Working Paper.

Engle, R. F., Ng, V., & Rothschild, M. (1990). Asset pricing with a factor ARCH covariance structure: Empirical estimates for treasury bills. *Journal of Econometrics, 45*, 213–237.

Ethier, N. E., & Kurtz, T. G. (1985). *Markov processes: Characterization and convergence*. New York: Wiley.

Evans, M., Hastings, N., & Peacock, B. (1993). *Statistical distributions*. New York: Wiley.

Feller, W. (1968). *An introduction to probability theory and its applications* (Vol. I). New York: Wiley.

Feller, W. (1971). *An introduction to probability theory and its applications* (Vol. II). New York: Wiley.

Fermanian, J. D., & Wegkamp, M. (2004). Time dependent copulas, Preprint INSEE, Paris, France.

Fréchet, M. (1935). Généralisations du théorè me des probabilités totales. *Fundamenta Mathematicae, 25*, 379–387.

Fréchet, M. (1951). Sur le Tableaux de Corrélation dont les Marges sont données, *Ann. Univ. Lyon*, 9, Sect. A, 53–77.

Fréchet, M. (1958). Remarques au sujet de la note préc édente. *Comptes Rendus de l'Academie des Sciences de Paris, 246*, 2719–2720.

Gagliardini, P., & Gouriéroux, C. (2007). An efficient nonparametric estimator for models with nonlinear dependence. *Journal of Econometrics, 137*, 189–229.

Gallant, A. R., & White, H. (1988). *A unified theory of estimation and inference for nonlinear dynamic models*. Oxford: Basil Blackwell.

Genest, C. (1987). Frank's family of bivariate distributions. *Biometrika, 74*, 549–555.

Genest, C., & Rivest, L. (1993). Statistical inference procedures for bivariate Archimedean copulas. *Journal of the American Statistical Association, 88*, 1034–1043.

Genest, C., Ghoudi, K., & Rivest, L. (1995). A semiparametric estimation procedure of dependence parameters in multivariate families of distributions. *Biometrika, 82*(3), 543–552.

Gibbons, J. D. (1992). *Nonparametric statistical inference*. New York: Marcel Dekker Inc.

Gourieroux, C., & Robert, C. Y. (2006). Stochastic unit root models. *Econometric Theory, 22*, 1052–1090.

Granger, C. J. W., Maasoumi, E., & Racine, J. (2004). A dependence metric for possibly nonlinear processes. *Journal of Time Series Analysis, 25*, 649–669.

Hansen, B. (1994). Autoregressive conditional density estimation. *International Economic Review, 35*, 705–730.

Huang, W., & Prokhorov, A. (2008). Goodness-of-fit test for copulas, Revised and Resubmitted, *Econometric Reviews*.

Ibragimov, R. (2005). Copula-based dependence characterization and modeling for time series, Harvard Institute of Economic Research, Discussion Paper No. 2094.

Nelsen, R. B. (1991). Copulas and association. In G. Dall'Aglio, S. Kotz, & G. Salinetti (Eds.), *Advances in probability distributions with given marginals* (pp. 51–74). Dordrecht: Kluwer Academic Publishers.

Nelsen, R. B. (1996). Nonparametric measures of multivariate association. *Distribution with fixed marginals and related topics* (pp. 223–232). IMS lecture notes-monograph series.

Nelson, C. R., & Plosser, C. I. (1992). Trends and random walks in macroeconomic time series. Some evidence and implications. *Journal of Monetary Economics, 10*(I 982), 139–162.

Politis, N., & Romano, J. P. (1994). The stationary bootstrap. *Journal of the American Statistical Association, 89*(428), 1303–1313.

Prokhorov, A., & Schmidt, P. (2009). Likelihood-based estimation in a panel setting: Robustness, redundancy and validity of copulas. *Journal of Econometrics, 153*, 93–104.

Réemillard, B., Papageorgiou, N., & Soustra, F. (2010). Dynamic copulas, Technical Report G-2010-18, Gerad. 1.

Rényi, A. (1959). On measures of dependence. *Acta Mathematica Academiae Scientiarum Hungarica, 10*, 441–451.

Robinson, P. (1983). Nonparametric estimators for time series. *Journal of Time Series Analysis, 4*, 185–207.

Rockinger, M., & Jondeau, E. (2001). Conditional dependency of financial series: An application of copulas, Les Cahiers de Recherche 723, HEC Paris.

Savu, C., & Trede, M. (2008). Goodness-of-fit tests for parametric families of Archimedean copulas. *Quantitative Finance, 8*, 109–116.

Scaillet, O. (2000). Nonparametric estimation of copulas for time series, IRES Working Paper.

Schmid, F., & Schmidt, R. (2006). Bootstrapping Spearman's multivariate rho. In A. Rizzi & M. Vichi (Eds.), *Proceedings of COMPSTAT* (pp. 759–766).

Schmitz, V. (2003). Copulas and stochastic processes, Ph.D. dissertation, Aachen University.

Sklar, A. (1973). Random variables, distribution functions, and copulas. *Kybernetica, 9*, 449–460.

Sklar, A. (1996). Random variables, distribution functions, and copulas - a personal look backward and forward. In L. Ruschendorf, B. Schweizer, & M. D. Taylor (Eds.), *Distributions with fixed marginals and related topics* (pp. 1–14). Hayward: Institute of Mathematical Statistics.

White, H. (1982). Maximum likelihood estimation of misspecified models. *Econometrica, 50*, 1–25.

White, H. (1984). *Asymptotic theory for econometricians*. New York: Academic Press.

Wooldridge, J. (1986). Asymptotic properties of econometric estimators, University of California, San Diego, Department of Economics, Doctoral Dissertation.

Printed in the United States
By Bookmasters